Managing the Wild

Managing the Wild

Stories of People and Plants and Tropical Forests

CHARLES M. PETERS

New York Botanical Garden

Yale University Press

New Haven & London

A co-publication of The New York Botanical Garden Press
and Yale University Press.

Yale University Press books may be purchased in quantity for educational, business, or promotional use. For information, please e-mail sales.press@yale.edu (U.S. office) or sales@yaleup.co.uk (U.K. office).

Set in Janson Roman type by Integrated Publishing Solutions.
Printed in the United States of America.

Library of Congress Control Number: 2017943418
ISBN: 978-0-300-22933-2 (hardcover : alk. paper)

A catalogue record for this book is available from the British Library.

This paper meets the requirements of ANSI/NISO Z39.48–1992
(Permanence of Paper).

10 9 8 7 6 5 4 3 2 1

In Memoriam

Charles Merideth Peters

1920–1974

Susan Daves Peters

1924–2010

Stories are one of the fundamental ways in which we understand the world. They are probably our best maps and models of the world—and we may yet come to learn that the reason for this is that stories are some of the basic constituents of the world.

—Robert Bringhurst, *The Tree of Meaning: Language, Mind and Ecology*, 2006

Contents

Preface

This is a book of stories about indigenous forest management written by a scientist who has spent thirty years working closely with rural communities in tropical forests. With few exceptions, the communities depend on a variety of forest plants for their livelihood. These may be timber trees, edible fruits, spiny vines, barks for paper, or even the leaf bases of certain agaves that can be fermented to produce an alcoholic beverage. The relationships developed between the people in the community and the plants in the forest are subtle, yet surprisingly complex, and in the narratives offered here I describe a dimension of tropical forest use that has rarely been presented. The conventional view in industrialized countries holds that native populations have been exploiting local forest resources with little regard for the future, cutting and burning large areas of tropical forest, and depleting innumerable tropical plant resources. But such actions represent only one aspect of the interaction of communities with local forests. In every tropical region of the world, communities also work to plant trees, select favorable genotypes, weed, thin, and manage forests, and do their best to control harvesting to conform to the rate at which a given resource is produced. These are the activities I focus on in this book.

As an ecologist and a forester, I have been most impressed during my fieldwork by how much villagers know about the properties and uses of local plants and the intricacies of what needs to be done to promote the regeneration and growth of a particular species in the forest. Their application of this knowledge, year after year for hundreds of years, has changed the structure and composition of tropical forests, but these interventions can be hard to discern. Because outsiders frequently do not recognize the imprint of local management—and because they are often unwilling to ask the people doing the managing—they might assume that nothing purposeful is being done to sustain local forests. I have had the good fortune to learn, and in some cases to be part of, the backstory of a number of positive interactions between a community and a tropical forest.

I have compiled these narratives here to offer a different perspective on what happens when human beings and tropical forest interact. Using examples from Central and South America, Southeast Asia, and Africa, I focus primarily on my collaborations with different communities to manage tropical forests sustainably, as well as my work documenting ways communities were already doing this. I conducted household surveys, ran inventory transects, counted and measured thousands of trees, quantified annual growth, and sampled an inestimable number of native fruits, but most of all I asked a lot of questions—and I listened. This is how I found most of my stories.

I wrote this book for people who would like to know more about how indigenous communities that have lived in a tropical forest for hundreds of years manage to do so without depleting their resource base; for those who are interested in what happens when colonists move into a new area of forest and wish to exploit it commercially; and for readers who are curious about the incredible variety of different fruits and fibers and timbers and oleo-resins and latexes and medicinal plants that are produced in tropical forests. I aim as well to show policy makers, resource managers, and conservation advocates the potential benefits of giving communities a more prominent role in the conservation of local tropical forests.

Forest dwellers have been exploiting plant resources in tropical forests at varying intensities for thousands of years; I have spent the past three decades looking into the nuances of this interaction. I have found that different communities do different things, that they do these things for their own reasons, and that some of their interactions with the forest are both skillful and sophisticated—beneficial to the community with minimal long-term impact on the forest. The current understanding that many in the industrial West hold of tropical forests seems to be largely defined by the tension created between the concepts of *wild* and *managed.* There is a tendency to think, "If it's wild, we should probably keep everybody out," and to deny the management they cannot recognize. I hope here to help resolve, or at least explore, this tension, and add useful new details to the world's understanding of forest use in the tropics.

Maps

The maps that follow show the places where I conducted projects focused on the use and management of tropical forests by local communities. Each map is keyed to the chapter or chapters in which these projects are discussed.

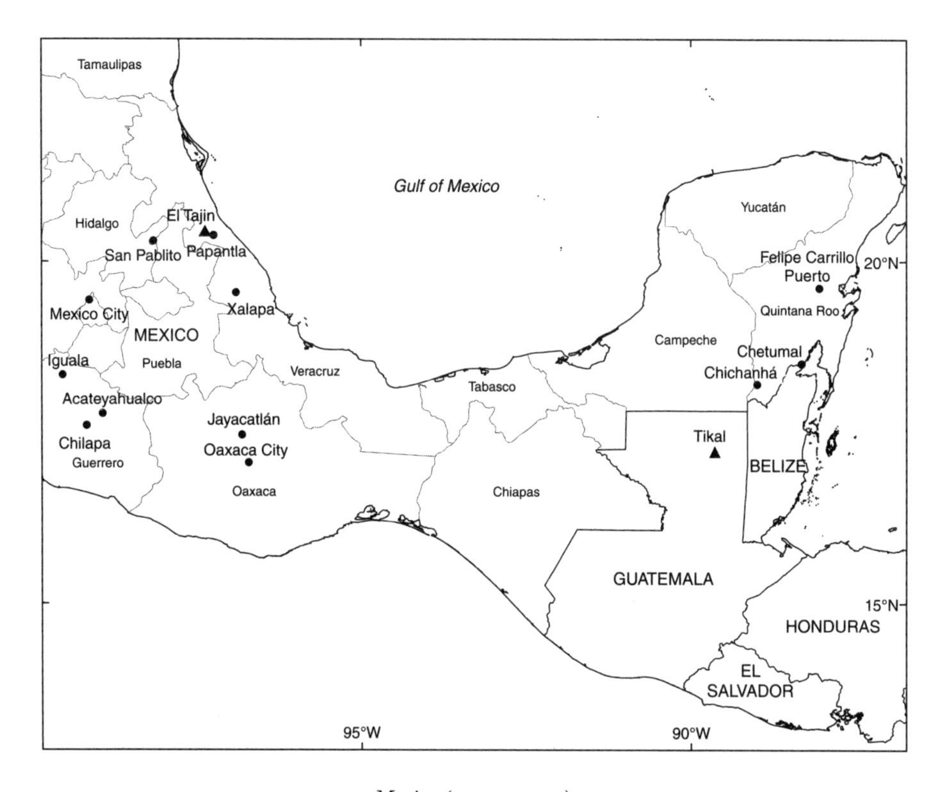

Mexico (1, 2, 9, 11, 12)

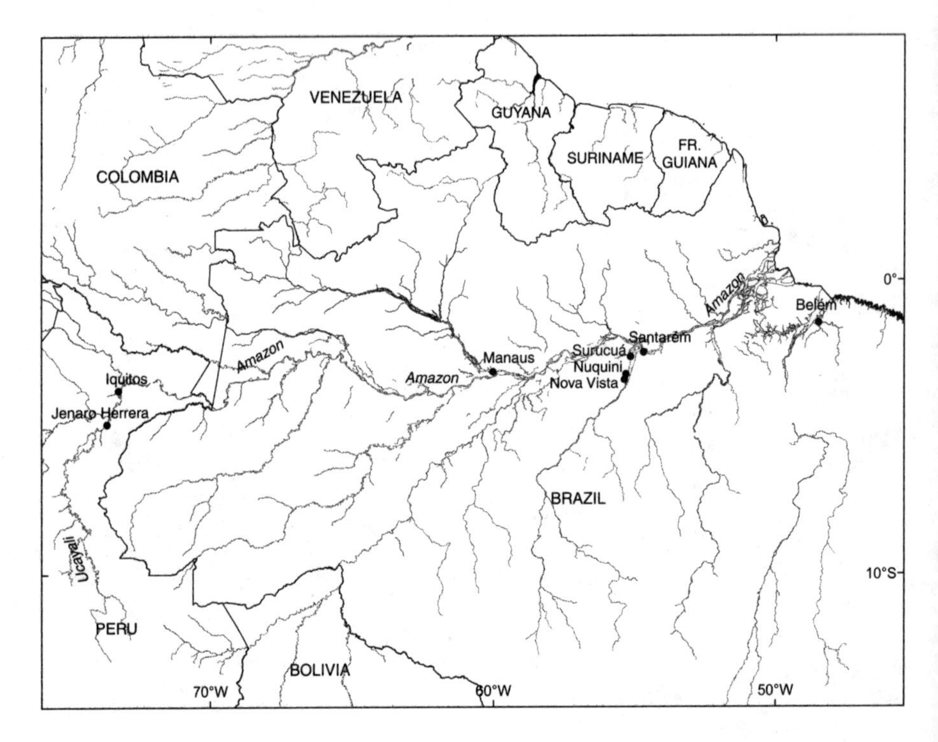

Amazonia (3, 4, 10)

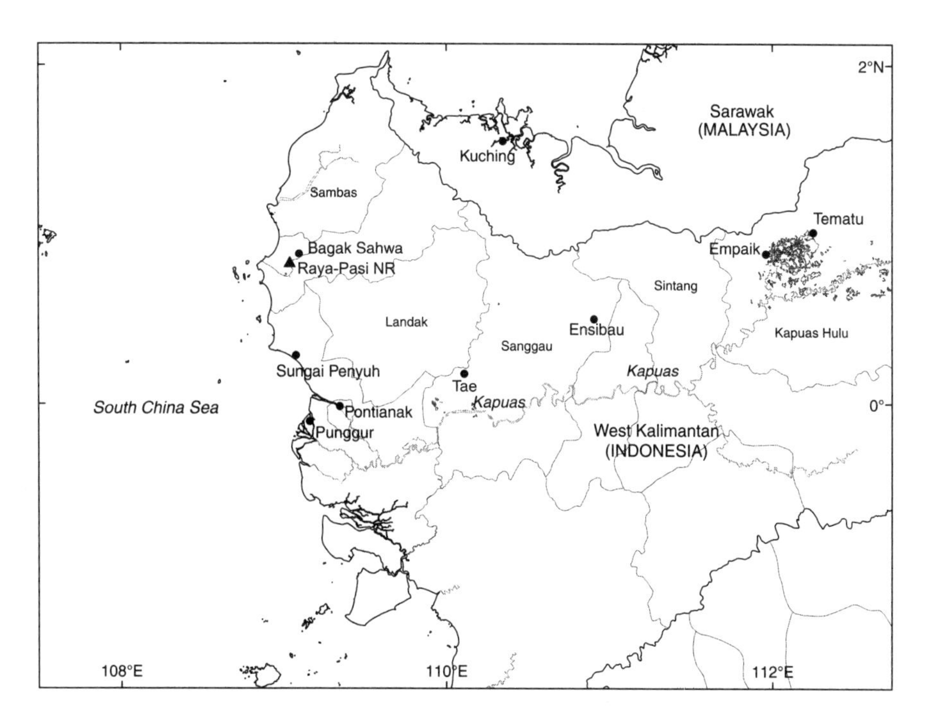

West Kalimantan, Indonesia (5, 6)

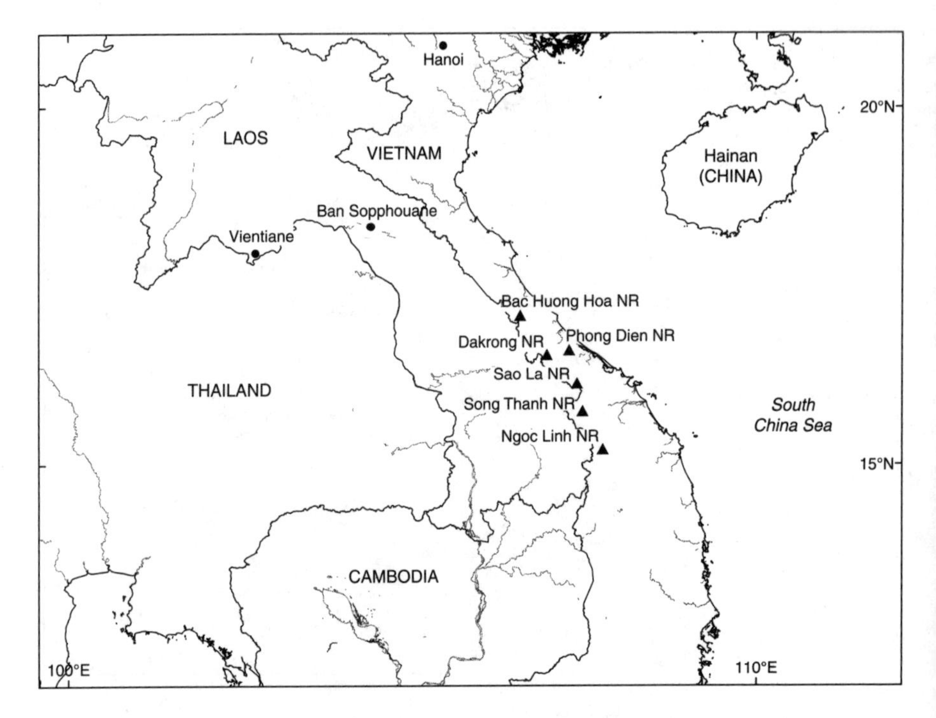

Greater Mekong Region, Laos, Cambodia, and Vietnam (14)

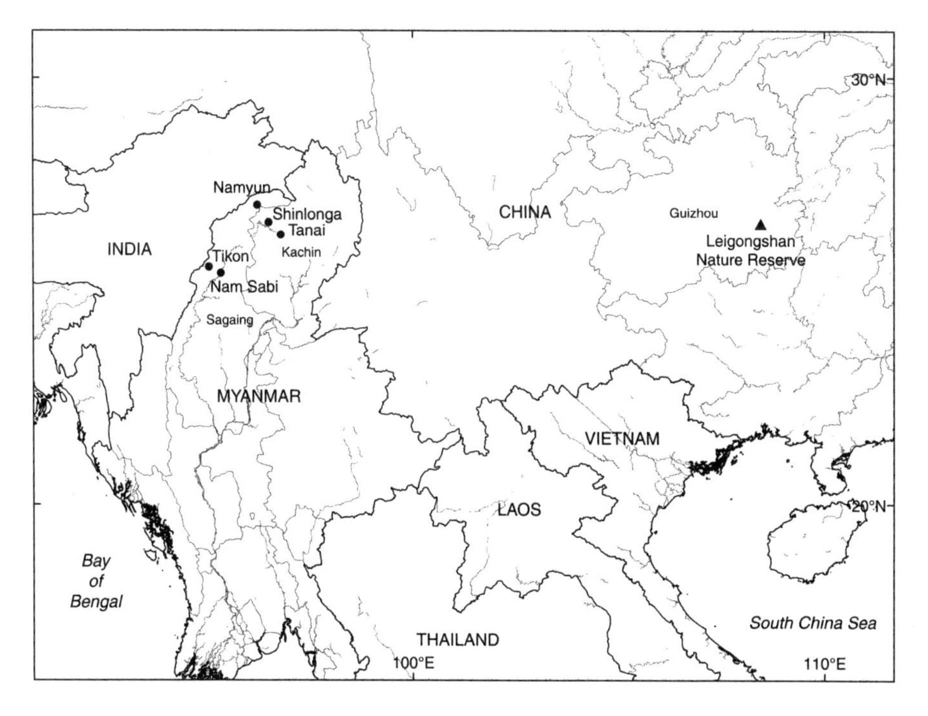

Southwest China and Myanmar (13, 15)

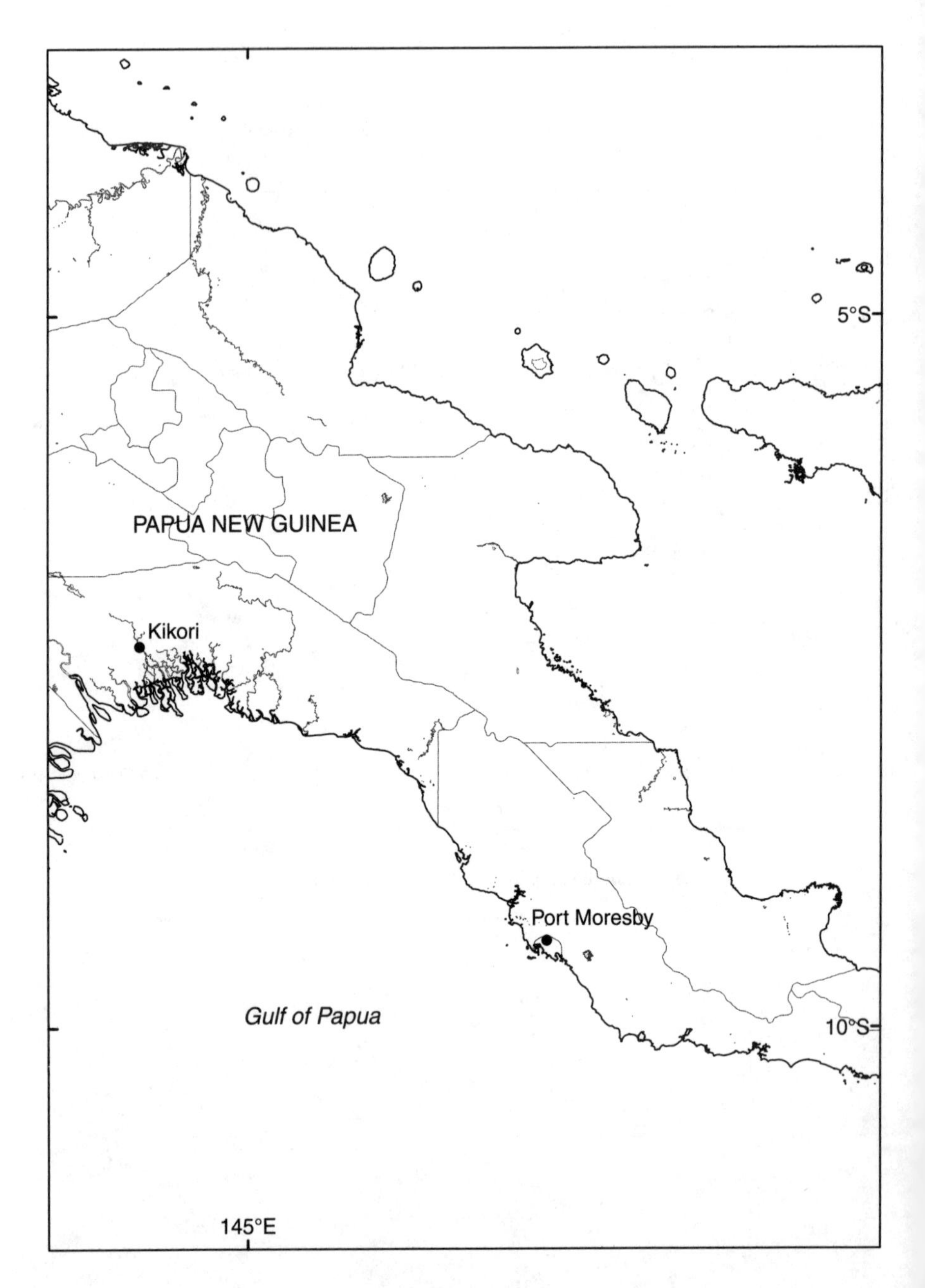

Papua New Guinea (7)

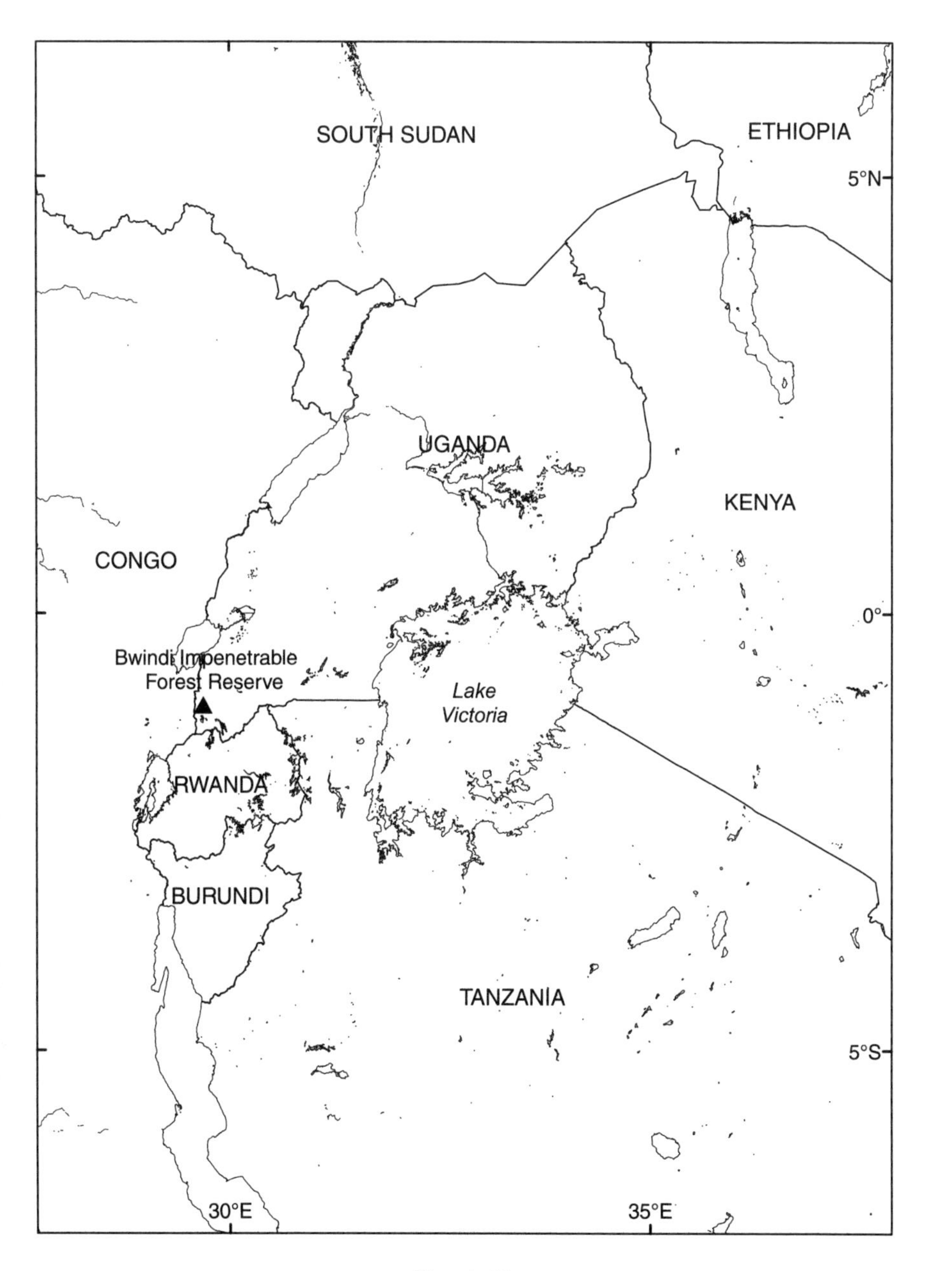
SOUTH SUDAN
ETHIOPIA
5°N
UGANDA
KENYA
CONGO
0°
Bwindi Impenetrable
Forest Reserve
Lake
Victoria
RWANDA
BURUNDI
TANZANIA
5°S
30°E
35°E

Uganda (8)

Managing the Wild

INTRODUCTION

The Challenge of Sustainable Forest Use

Tropical forests are best known for the wide variety of plants that grow in them. Small tracts of forest may contain hundreds of plant species; neighboring tracts may exhibit similar levels of floristic diversity, but the species are different. Given such a large assortment, it is not surprising that many of these plants have properties that have proven useful to human beings. Certain trees are good for boards or making canoes or framing houses. Some trees produce delicious fruit, aromatic spices, caffeine-rich leaves, or a valuable latex or resin.[1] The leaves of many palm species are used for thatch, while other species produce edible fruits or floral nectars that can be distilled into wine. The stem fibers from one group, the rattans, form the basis of a multibillion-dollar furniture industry.[2] Tropical forests contain plants that are used for medicine, plants that are used as contraceptives, and plants that are used as intoxicants. All these botanical resources have been used by indigenous communities for thousands of years. The main reason we know as much as we do about the uses of tropical plants is that forest dwellers have explained them to us.

Interactions between communities and tropical forests are not always positive, however. Resources can be overexploited or destructively harvested; small tracts of forest can be cut and burned for agriculture. These occurrences become more prevalent as population density and market involvement increase. More families need more land to grow food, and growing markets demand more timber, more fruit, or more rattan, regardless, it seems, of ecological impact. The general perception by many in the international conservation community holds that the best way to ensure the preservation of tropical forests is to get the people out of them. Stop the harvest of forest products, gazette more protected areas, hire more forest guards. The growing appreciation of the importance of tropical forests in ameliorating impending climate change seems to have strengthened the global mandate to separate the people from the forest.

While an extensive network of protected areas clearly has value for the conservation of biodiversity in the tropics, it is not the only way to approach the problem.[3] It is important to remember that the constant manipulation of tropical forests by human beings over time has created and shaped the plant communities that are currently being designated as parks and reserves. In many cases, human activity has actually increased the diversity in tropical forests by creating new habitats, selective weeding, and the introduction of new species. Once these human activities are constrained or eliminated, the floristic composition of the forest will surely change—and we might not like the result.

In addition, the exploitation of plant resources in tropical forests need not inevitably lead to resource depletion and forest degradation. Resource depletion happens only when the amount of plant material harvested exceeds the amount that the forest can reliably produce each year. This is the basic premise of sustainable resource management. If we know the existing *stock* of the resource and can obtain estimates—even crude estimates—of its *yield* each year (how much timber or resin or palm thatch is produced), we can estimate a sustainable level of harvest for the species. If, and this is a big *if*, harvest levels are respected and the community consistently collects less than the annual growth of the resource each year, that resource can be exploited year after year with minimal impact on the plant population that produces it.

The difficulty comes in trying to conduct the inventories to quantify the stocks of different species and in setting up the yield studies to estimate how much of a given resource the forest produces each year. This work has to be done in close collaboration with local communities so that they will learn how to collect the necessary data themselves, can appreciate how the data fit together to define a sustainable level of exploitation, and, most important, will develop the motivation to control how much they harvest each year. The temptation to depart from the onerous prescriptions of a forest management plan is ever present, especially in times of financial hardship or when market prices soar.

For the past thirty years, I have been involved in a variety of projects focused on the conservation and sustainable use of tropical forests by local communities. Some of the projects were successful; most provided valuable data and training in resource management; and a few turned out to be no more than temporary distractions from more critical issues a community was facing. The results from each of these investigations, however, can teach us something important about the current realities of resource use in the tropics.

Several factors motivated me to write a book about these field experiences. First, I have learned intimate details about the ecology, use, and management of a number of little-known yet potentially valuable tropical forest resources and am convinced that these data

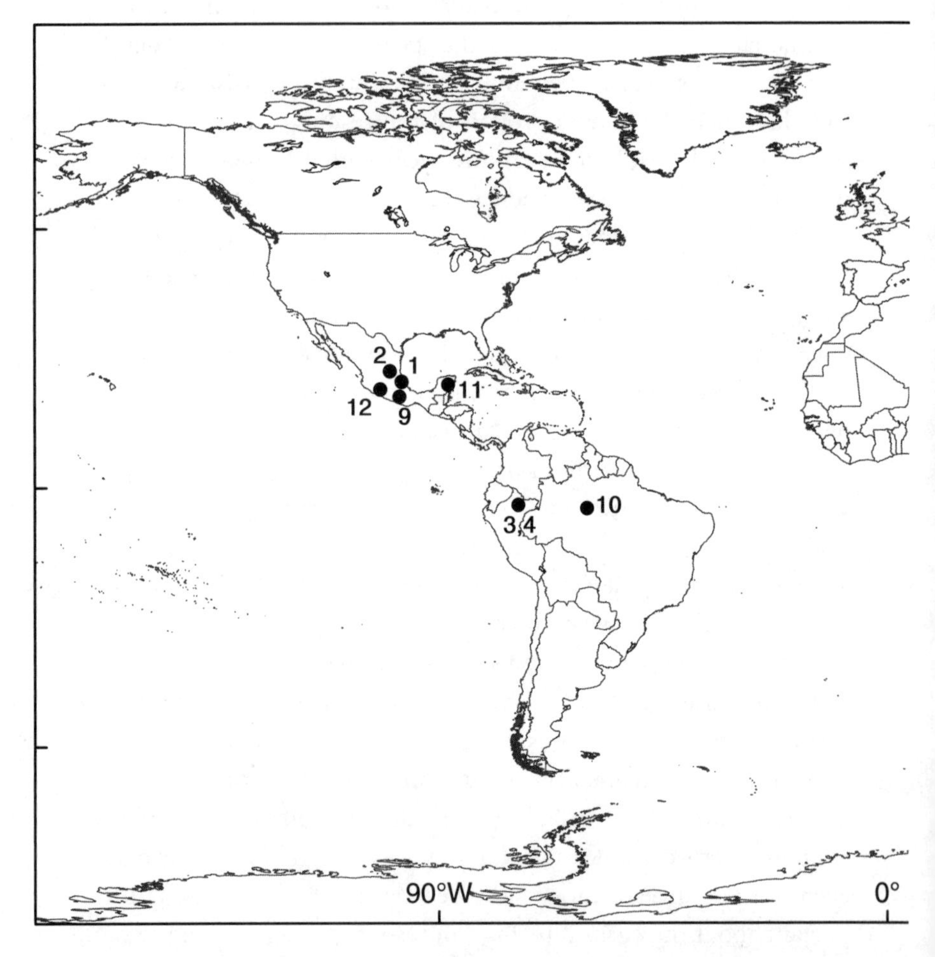

Location of projects I have been involved with that focus on the sustainable use of tropical forest resources by local communities. The numbers refer to the relevant chapter in this book; the dates of fieldwork are included on the first page of each chapter, and geographic coordinates of specific localities are also provided where applicable.

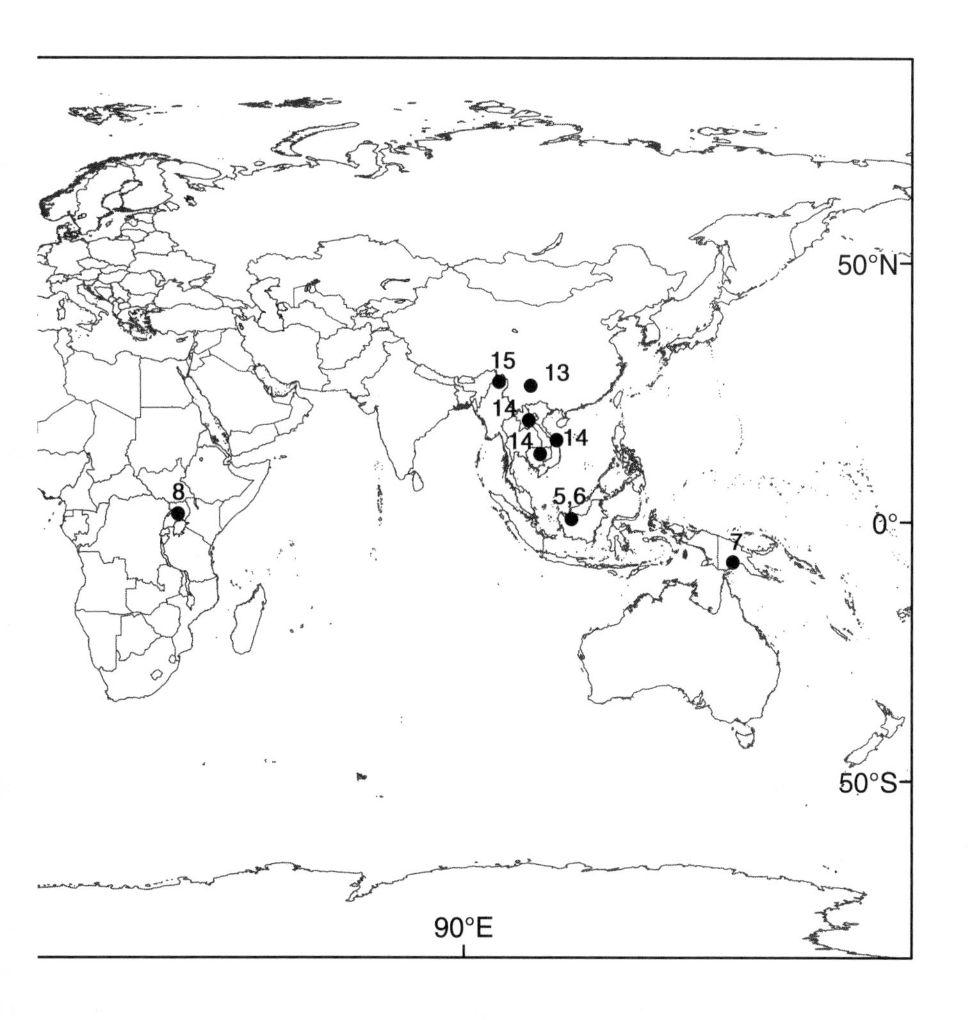

15
13
14
14
14
8
5,6
7
50°N
0°
50°S
90°E

could be put to good use by resource managers. In addition, the participatory protocols that were developed for managing wild populations of tropical plants offer valuable alternatives to deforestation. It is possible, and not overly difficult, to use tropical forests in a way that both conserves the forest and provides local communities with an incentive for forest stewardship. There is an attractive logic associated with helping communities develop sustainable management systems for their forests, rather than excluding them from decisions that directly affect them or paying them to leave the forest alone.

I have had the good fortune to work with forest-dwelling communities in Asia, Central and South America, and Africa, attempting to do, more or less, the same type of project with various resources, ethnic groups, and forest types. Such a replication of field methods across so many different contexts is rare, and I thought it worthwhile to bring all the projects together—the successful ones, as well as the less successful—for comparison. Many of the experiences I have collected during my fieldwork are at odds with some of the things people usually hear about tropical forests. For instance, none of the forests that I worked in, even in the most remote locations in Amazonia, Borneo, and Papua New Guinea, were "pristine" or untouched. Local communities had been living in these forests, farming, planting trees, and managing the vegetation, for centuries. But to learn what indigenous peoples have done, we have to ask them because the imprints of their management activities are, in many cases, essentially invisible—something that for me constitutes one of the hallmarks of successful silvicultural treatment.

I have published many scientific papers about this work, but the present offering is not a scientific monograph. Each chapter, or case study, offers a narrative that illustrates different aspects of a particular forest resource, community group, and interaction between people and plants. The stories illustrate the ecological conditions or social contexts I encountered when I started my work, describe the intervention or data collection process I facilitated, and help me summarize the results.

The level of detail provided, much like my field notes and recollections, varies from chapter to chapter depending on the place, my skill with the local language, and (though I hate to admit it) how long ago I did the work. (The chapters are ordered chronologically.) Some of these stories offer snapshots of ecological understanding, careful resource use, and indigenous silviculture; a few describe difficult situations in which neither the people nor the plants seem to be thriving. It is my hope that these stories will help future researchers find more effective and equitable strategies for the long-term conservation of tropical forests. All the projects, in hindsight, have involved delightful people, lots of laughter, long conversations about plants, and mostly fruitful if largely unexpected outcomes. And this, perhaps, is what I would most like to share with the reader—the reward of thirty years of concerted trial and error.

ONE

The Ramón Tree and the Maya

Twelve kilometers southwest of Papantla (N20°26′52″, W97°19′12″), Veracruz, Mexico, 1979–1983

A number of Maya ruin complexes in Mexico and Guatemala were discovered by *chicleros* looking for forage to feed their mules.[1] The preferred forage was the leaves and twigs of a common forest tree known locally as *ramón* (*Brosimum alicastrum*),[2] from the Spanish verb *ramónear*, "to browse." The chicleros would encounter a dense stand of ramón trees, start lopping off the branches, and eventually notice that the trees were growing on top of the stones and broken statuary of a Maya temple.

The relationship between ramón trees and Maya ruins can tell us much, given the many uses of the tree.[3] The leaves are used for forage, the fruits and seeds are edible, the milky white latex is drink-

(*Opposite*) *Ramón* (*Brosimum alicastrum*) trees growing on the ruins of Coba in Quintana Roo, Mexico.

able, and the wood is durable and easily worked. The dried ground seeds contain more protein than corn and have a high concentration of tryptophan, an essential amino acid that is usually deficient in diets based mainly on corn. Rural communities throughout Central America currently eat ramón seeds as a survival food, and a sizable archeological literature suggests that the seeds were a dietary staple of the Maya in pre-Columbian times. Many investigators conclude from this that the dense aggregations of ramón trees growing on ruin sites are, in essence, relict Maya orchards.

I first found out about the ramón tree when I was looking for a research topic for my Ph.D. dissertation. I had read a little bit about the Maya–ramón tree connection, but I was more interested in studying the population ecology of a tropical tree that was able to form high-density aggregations. Tropical forests are so diverse because they have a wide variety of niches and growth strategies, and all the plant species behave in subtly different ways to survive in a highly competitive environment where one species is rarely able to function so much better than all the others that it can take over. So the fact that ramón trees were, apparently, able to achieve such a notable degree of dominance—on top of Maya ruins or anywhere else—was a finding of great interest to a fledgling tropical ecologist. I really wanted to figure out how the species was able to do this.

A research institute in Xalapa, Veracruz (Instituto de Investigaciones sobre Recursos Bióticos A.C.), had agreed to host me during my doctoral work, and I spent my first months in Mexico traveling around the state of Veracruz visiting different forests and counting ramón trees. I was trying to locate the perfect study site for my research. The forests in the southern part of the state contained about four to five adult ramón trees per hectare, while the ones in the far north, in drier, cooler climates along the border with the states of Tamaulipas and San Luis Potosi, had ramón populations of over one hundred adult trees per hectare. None of these sites met my needs. The densities of the ramón populations in the south were too low for my purposes, and the sites to the north, while chock-full of ramón trees, were too far away from Xalapa.

The forest I finally chose to work in was located about three hours north of Xalapa near the town of Papantla and the ruins of El Tajin, a complex of Totanac pyramids built between 600 and 1200 C.E.[4] The forest was a beauty: huge trees with sinewy buttresses, palms and delicate ferns in the understory, flocks of parrots and the occasional toucanet, and, in terms of the ramón population, thirty to forty adult trees, numerous poles and saplings, and thousands of seedlings per hectare. The species was clearly happy growing here. This was the first tropical forest that I spent a great deal of time in, collecting and counting and thinking. The locals called the site Los Alpes because it was incredibly steep, but the climb did not bother me.

The basic focus of my work at Los Alpes was to document the birth, death, and growth rates of all the ramón trees. The procedure is called demography and is similar to what insurance companies and census bureaus do to calculate the probability that a person will die at a certain age or to estimate the size, age structure, and potential growth rate of a population. I wanted to use the tools of ecology and plant demography to see if the ramón trees would be able to maintain their dominance in the forest. Were these populations dense because they had been planted and managed, or were they dense because the life-cycle characteristics of the ramón tree were so effective in this habitat that they did not need humans to help them?

I carefully laid out a one-hectare study area in the forest. I identified every tree, shrub, and herbaceous plant, measured the diameter and height of all of the trees, and gave each a little round metal tag with a unique number. There were too many seedlings to count, so I made regeneration plots in random locations throughout the forest to sample the smaller-size classes. I put stainless-steel growth bands around all the ramón trees, and around several representatives of each of the associated tree species. Growth bands, also known as dendrometer bands, are custom fit around the trunk of a tree, and kept in place by a spring. The installer marks a vertical "zero" line, and as the tree grows the upper and lower sections of the line separate to indicate diameter growth; the distance between the two lines is measured with a vernier caliper. Dendrometer bands are extremely sensitive, and once the tree has grown sufficiently into the band, the swelling and shrinking of the bark after a rainstorm is clearly visible.[5] Finally, I built several dozen litter traps, using wooden stakes, hoops of plastic tubing, thick garbage bags, and clothespins, and set them under the crowns of selected ramón trees to catch the flowers and fruit that fell.

With these pieces in place, all I had to do was return to the site every few weeks. I needed to count the number of seedlings in my plots; note any trees that had died and record, if possible, the reason they had died; read the growth bands; and, if the ramón trees were flowering or fruiting, change the garbage bags in the litter traps and carry them back to Xalapa so I could sift through the contents and count the flowers and fruit. Although tedious, the work was straightforward. If done correctly, by the end of one year, I would have a reasonable estimate of the birthrate, the death rate, and the growth rate for all the trees in the population. But I admit that going up and down the steep slopes with field equipment and garbage bags full of fruit did not get any easier over the course of a year.

One evening, my truck broke down (it needed a new generator), and I had to take a bus back to Xalapa. As it happened, I had just emptied all my litter traps, and I needed to carry the contents back with me. I never knew what I was going to find when I stuck my hand down in those traps. Snakes seemed particularly drawn to the moist, dark space, and I encountered black scorpions a couple of times as well. What was guaranteed, however, was that all the flowers, fruit, and leaves, and everything else that had fallen into the trap, would have turned into a wet, stinking mess with maggots crawling through it. I usually threw the garbage bags into the back of the truck and picked through them wearing latex gloves when I got back to Xalapa. But this time, I had to carry the smelly mass of plant parts on a crowded public bus.

It was almost dark by the time I caught my bus, and it was packed—señoras with long braids wearing embroidered blouses, gunny sacks full of beans and corn, a couple of wooden crates with pigs, chickens flailing around, and a group of little kids. Every seat was filled. I got some curious looks as I stepped in with my black garbage bags, but I slowly made my way to the back of the bus and found a place to stand. The ride took four hours, and after about thirty minutes the bus was filled with the pungent odor of rotting fruit. Everyone was very tolerant of the stench and I was even offered a seat near the end of the trip.

I continued to monitor flower and fruit production by the ramón trees at Los Alpes for two years and discovered some interesting things. For example, a ramón tree can be female, hermaphroditic, or male.[6] All the small trees were female, most of the intermediate-size trees were hermaphrodites, and all the large trees were male. I did not understand why this should be until the second year of fruit collection. One of the female trees I had monitored the year before, which had produced only female flowers and fruit, started producing male flowers. Similarly, one of the hermaphroditic trees, which had produced male and female flowers and a few fruits the year before, produced only male flowers and no fruit; functionally, it had become a male tree. Changing sex is not common in tropical trees, so far as we know, but the ramón trees were exhibiting what is called "sequential hermaphroditism"—they can modify their sexual expression if necessary.

Based on the results of the growth-band studies, I deduced that the benefit of changing sex from female to hermaphroditic to male is to allow a tree to grow faster and outcompete its neighbors. Female trees are forced to mature tens of thousands of protein-rich fruits, while male trees immediately go back to growing after they release their pollen. If the canopy conditions above a female tree remain favorable, the tree will continue to produce fruit. If, however, light levels drop too low or the competition from neighboring trees becomes too intense, the tree will start producing male flowers. In terms of its reproductive biology, the tree is now hermaphroditic. It will still produce a few fruits, but its growth rate will be significantly higher than that of a female tree. If canopy conditions or competition levels continue to be restrictive, the tree will stop producing female flowers altogether. I would guess that this type of plastic sexual expression is more common in tropical trees than most ecologists realize.

Another curious aspect of the reproductive dynamics of ramón trees is the role frugivorous bats play in seed dispersal. A number of animals, including humans, eat ramón fruit in large quantities. I have watched flocks of parrots feed for hours in the crowns of fruit-laden ramón trees, and a number of other types of birds, as well as monkeys, squirrels, and raccoons, also relish it. But these animals eat the fruit, and the seeds are either destroyed in the process or fall to the ground directly under the crown of the tree. Because of the excessive competition with siblings—other seeds produced by the same tree—this is not an ideal place for a seed to be deposited. Frugivorous bats, on the other hand, will fly to a fruiting tree, select one fruit, hold it carefully with their feet, and then fly away to a roost, where they will carefully eat the fruit and then drop the seed intact, a textbook example of successful seed dispersal.

Large colonies of bats live in the inner chambers and vaults of Maya temples, and the floors and limestone ledges of these ruins are usually littered with ramón seeds. I have seen extensive accumulations of ramón seeds and seedlings around the ruins of Palenque and Bonampak in Chiapas, and under bat roosts in mango plantations in Veracruz, where the nearest ramón tree was more than five kilometers away. I have a hunch that the commonly observed aggregation of ramón trees around Mayan ruins is at least partly the result of the inherent competitive ability of the species coupled with a continual input of bat-dispersed seeds.[7]

Of course, accepting the hypothesis that the clustering of ramón trees on ruins is the result of bat dispersal and normal ecological processes does not negate the possibility that the tree was used—or even managed—by the Maya. It is hard to believe that such an abundant, protein-rich food source would not be exploited in some manner. The issue here appears to be one of the extent or magnitude of use. Were ramón seeds casually collected from the forest only in times of famine, or was some form of deliberate cultivation, silviculture, or selection practiced?

I wondered whether evidence of intense use by the pre-Columbian Maya might still be detectable in the ramón populations growing in the forests around some of the major population centers, like Tikal, one of the civilization's largest archaeological sites and urban centers.[8] Much of the work on the historical use of ramón by the Maya has focused on Tikal, and I thought it might be productive to compare these data to the ecological information that I was collecting at Los Alpes. What I found was striking.

The ramón trees at Los Alpes—and in most of the other places where the species grows—produce fruit once a year, with peak fruiting occurring at the onset of the rainy season. This makes biological sense. Flowers are pollinated during the dry season, and the seeds have a moist substrate for germination and early growth. Ramón trees are pollinated by wind, a poor carrier of pollen compared to bees, beetles, or bats, especially when it is raining, and this form of abiotic pollination is rare among tropical trees. By contrast, reports from Tikal suggest that the ramón trees there bear fruit twice a year or that fruiting is largely continuous throughout the year, with three periods of peak fruit abundance. It is worth noting that the environmental conditions at Tikal—elevation, temperature, precipitation, substrate—are virtually identical to those at Los Alpes.

The reproductive behavior, or breeding system, of the ramón trees at Tikal is also different from those at Los Alpes. In contrast to the complicated sequential hermaphroditism and facultative sex change that I found at Los Alpes, the ramón trees at Tikal are consistently described as having both male and female flowers, and all the trees are capable of producing fruit.[9]

What we seem to be presented with at Tikal is a dense population of ramón trees in which fruit is produced in large quantities at frequent intervals by every adult tree. The curious feature is that similar forests of the species in other regions do not exhibit this behavior. One explanation might be that the regeneration of the species is severely limited for some reason at Tikal, and this large reproductive output is required to maintain the population. Another explanation, which I find more convincing, is that the high level of fruit production exceeds the regeneration needs of the population, and the atypical reproductive dynamics of the ramón trees at Tikal are manifestations of relict genotypes selected for by the continued use and management of the tree. Excessive collection of the seeds during times of famine would not produce such a marked effect. What appears to be suggested here is the conscious mixing of genotypes with the objective of producing an abundant, year-round supply of seeds.

At this point in my career, I was intrigued by questions of theoretical ecology and niche theory, and spent a lot of time collecting data to either prove or refute current theories about the population dynamics of tropical trees.[10] I spent more time obsessing over what my data might tell me than reflecting on the motives and expertise of the people who had, in all probability, subtly groomed the floristic composition of the "undisturbed" piece of tropical forest in which I was doing my doctoral research. But my focus was starting to change.

The insight that the Maya had deliberately managed their forests during pre-Columbian times to increase the density and yield of ramón trees inevitably gave rise to the question of whether indigenous groups of tropical forest dwellers were doing the same thing today. If so, how were they going about this? If not, could some of the skills that I had learned as a forester and an ecologist be applied within a community context to facilitate this type of forest management? My fieldwork after the ramón study became more focused on practical applications. I began to put together projects that were collaborative efforts with local communities aimed at solving a problem rather than proving a theory. The nature of the stories I heard, and gradually became a part of, was also starting to change.

TWO

Mexican Bark Paper

Commercialization of a Pre-Hispanic Technology

San Pablito (N20°18′02″, W98°09′45″) in the Sierra Norte of Puebla, Mexico, 1984

I continued working at the research institute in Xalapa after I finished my doctoral research. One of my first projects, which represented a notable shift from my ecological work on the ramón tree, was focused on the bark paper trade in Mexico. These brightly painted sheets of handmade paper had long been a successful handicraft item for tourists, but several aspects of the bark paper supply chain had changed dramatically in recent years. My research uncovered a fascinating history of people and plant interactions, and provided a sobering example of how market conditions and poor forest management can lead to resource depletion.

(*Opposite*) Otomi woman using strips of *jonote* (*Trema micrantha*) bark to make a sheet of paper in San Pablito, Puebla, Mexico. Note the softened bark fibers hanging in the background, and the rectangular, flattened stone, or *muito*, used for pounding the bark fibers together.

Paper made from tree bark has been produced in various parts of Mexico for over fourteen hundred years. Bark paper played an important cultural role during pre-Hispanic times, and many indigenous groups developed papermaking technologies. The total quantity of bark paper produced in Mexico before the arrival of the Spanish was sizable. In the *Codex Mendoza*, a history of the Aztec rulers with a list of the tributes offered them each year that was written after the Spanish conquest, it was reported that over 480,000 sheets of bark paper were sent in annual tribute to the royal house of Moctezuma II in Tenochtitlán, the capital city (today's Mexico City).[1] After arriving at the imperial city, the paper was destined for use in religious ceremonies, to adorn temples and altars, and for codices. The paper appears to have been produced by roughly fifty villages scattered throughout the Aztec Empire.

The Otomi people first moved into the rugged mountains of northern Puebla around 800 C.E. and were conquered by the Aztecs in the late sixteenth century. One of the local Otomi communities, San Pablito, had an active tradition of papermaking, and this town was undoubtedly a major producer of the bark paper given in tribute each year to Moctezuma. After the Spanish conquest, the production of bark paper in Mexico gradually started to disappear. San Pablito was one of the few communities that continued to make the paper for use in agricultural rites, folk medicine, and witchcraft, although it is doubtful that large quantities were required to satisfy these ceremonial needs. The papermakers in San Pablito continued at a low level of production for several centuries.

But in the early 1960s things changed. An architect, Max Kerlow, and a painter, Felipe Ehrenberg, gave several sheets of Otomi bark paper to native artists in Guerrero renowned for their intricate, colorful paintings on clay pots and wooden masks. The painted bark paper sheets were received with enthusiasm, and in 1963 the first exhibition of painted bark paper was held in Mexico City.

The exhibition was a great success, and the demand for bark paper paintings increased dramatically.

Today, bark paper paintings represent one of the most successful and widely distributed types of folk art in Mexico. They are sold in tourist centers throughout the country and exported in large quantities to American, European, and Japanese markets. The production necessary to satisfy the commercial demand for bark paper paintings exceeds a hundred thousand sheets in some months—more than twice the amount sent in annual tribute to Moctezuma II. The salient difference is that San Pablito is currently the only remaining papermaking center in Mexico, and this small Otomi village makes all the bark paper required to supply the enormous market. The drastic increase in demand over the past fifty years has produced several notable changes in the way bark paper is made in San Pablito.[2]

San Pablito is located deep in the mountains at the end of a steep and winding road. The town is very quiet. All that can be heard from most of the houses is the constant tapping of a rock on a board as bark fibers are flattened and fused together to make paper, even during the middle of the day, when everyone might be expected to be working in the fields. Not many young men can be seen around the village. Although the town seems to be doing well—it has a high school, electricity, telephones, and computers—all the younger men have gone to the United States to find work, leaving the women and children to pound out the bark paper and negotiate the contracts with the buyers.

The Spanish phrase for bark paper, *papel amate*, identifies it as paper (*papel*) made from the bark of fig (*amate*) trees, trees from the genus *Ficus* (Moraceae). The inner bark of fig trees provides the perfect raw material for making paper. The fibers are stringy and flexible and they can be easily separated after they have been boiled for a couple of hours. Once they are softened, they can be pounded together on a board with a flat rock to form a sheet of interlacing fibers that, after it dries in the sun, turns into coarse paper. Fig trees have the added advantage of being able to regenerate bark after it is stripped off. A single fig tree can be exploited repeatedly for bark fiber—as long as not all the bark is harvested at once and the tree is given sufficient time to recover between harvests.

Traditionally, about half a dozen different species of fig tree were used to make bark paper. Although fig trees are common components of tropical lowland forests, the forests surrounding San Pablito represent a transition between the tropical lowlands and the temperate highlands. The climate is humid, yet temperate, with a marked seasonality of rainfall and a pronounced drop in temperature at night. Fig trees do not generally thrive in this type of climate, and local forests contained only a few scattered individuals from the genus. As long as the use of paper in San Pablito was limited to witchcraft and other ceremonies, a sufficient amount of bark fiber could be obtained from these trees. When the demand for bark paper began to increase, however, the Otomi were forced to remove larger strips of bark and increase the frequency of collection. The bark was not allowed to regenerate, entire trunks were denuded, and all the fig trees in the forest eventually died.

During this period, the expansion of agriculture in the region prompted forest clearing so the inhabitants could plant coffee and citrus trees; pastures were also established and livestock introduced. Areas formerly in forest were soon covered with communities of fast-growing pioneer species. One of the most common weed trees found in these early successional thickets was *jonote* (*Trema micrantha*),[3] and in the early 1980s the Otomi in San Pablito started making paper from this species. Although the bark fibers of jonote are thicker and more rigid than those of fig trees, longer boiling times and the addition of lime (calcium oxide) to the water were introduced to soften the fibers. In its favor, jonote occurs in dense aggregations, not as scattered individuals, and grows extremely fast. The bark can also be harvested throughout the year, in contrast to fig trees, which must be harvested at the onset of the rainy season to facilitate bark removal. The disadvantage of jonote is that the species does not regenerate its bark. A section of stem can be harvested only once, and if excessively long strips of bark are removed, the tree dies. The need for ever increasing amounts of bark fiber soon killed all the jonote trees in the vicinity of San Pablito.

The paper currently produced in San Pablito is made from jonote bark brought in by truck from the state of Veracruz, over 250 kilometers away. While a stack of painted bark papers for sale in a craft shop or the Mexico City airport may not seem noteworthy, an ordinary sheet of Otomi bark paper follows a long, complicated route before arriving at its final destination. The raw material is shipped from Veracruz, the paper is manufactured in Puebla, and the final painting is done in Guerrero. And perhaps more to the point, there is no amate in this papel amate.

Trees produce two types of cells: *meristematic cells*, which can divide and produce more cells, and *structural cells*, which provide support and transport water and nutrients to different parts of the tree. A single layer of meristematic cells, the vascular cambium, circles the diameter of a tree. As the cells in this layer divide, they produce one new type of cell, the xylem, toward the inside that conducts water and nutrients upward, and another new type of cell, the phloem, toward the outside that conducts the photosynthetic products from the leaves down to the rest of the tree. Given the volume of material that needs to be transported, the xylem cells, the cells that pump water, are produced in much greater quantity than the phloem. When bark is stripped from a tree, the tissue containing these cells is removed, and in some cases the single layer of meristematic cells also comes off. Without meristematic cells, the tree cannot regenerate its bark. If all the bark is removed from the stem of a tree, the tree will eventually die because the plumbing that supplies the sugars needed for growth has been removed.

The apparent inability to regenerate its bark was a major drawback in the use of jonote trees. I wondered whether protecting the bare trunk from desiccation after harvesting the bark might help stimulate the wound-healing process. If jonote trees were able to regenerate their bark, collectors could harvest them more than once, leaving a few sections of bark intact along the stem each time, so the harvest tree would not die. The bark resource could be managed on a sustainable basis, a new generation of jonote trees could be planted near San Pablito, and the critical link between artisan and raw material could be reestablished. The key was finding a way to promote the production of new bark tissue.

To test this idea, I located a dense stand of jonote trees near my office in Xalapa, stripped varying percentages of bark from several trees and then wrapped the trunks of some of the individuals in banana leaves and others in aluminum foil.[4] Anatomical analyses of samples of the bark and wood from the harvest trees indicated that, in every case, I had ripped off all the meristematic cells when I stripped the bark from the jonote trees.[5] The bare, exposed trunk, which I carefully wrapped to keep it moist, was composed entirely of structural cells.

Every two weeks, I would briefly uncover the trunks of the trees and take another wood sample. For the first six weeks, nothing happened. The tissue remained moist, but the wound that I created when I stripped off the bark had not begun to heal. During the next sample period, however, new tissue started to form. A layer of structural cells had divided and was producing new water-conducting xylem cells to the inside, and new photosynthate-conducting phloem cells to the outside. And new bark was growing along the edges of the wound. At the time, this seemed like magic.

An additional objective of my work at San Pablito was to try to reestablish a local source of bark fiber for the Otomi artisans. I had initially hoped that silvicultural techniques could be used to enhance the regeneration of the original fig tree populations in the region, but the scarcity and poor condition of existing trees made this plan untenable. After discussions with the villagers, it was decided to establish a plantation of jonote trees. This species is easy to propagate, attains a harvestable size in four to five years, and is well adapted to the high light conditions of a plantation; with careful post-harvesting treatment it will also regenerate its bark. The villagers would still have the option of planting fig trees, which are more shade tolerant, under the jonote trees after the canopy of the plantation had started to close. A nursery containing ten thousand jonote seedlings was established on the outskirts of San Pablito in 1984, and a local farmer donated a small tract of land to transplant the seedlings to when they reached sufficient size.

I went to Peru shortly after the nursery was established and have never revisited San Pablito. I am told that the production of bark paper continues to increase. New products like bark paper envelopes, lampshades, wallpaper, and faux parquet flooring have been developed, leading to chronic shortages of raw material. Apparently, nothing was ever done with the jonote seedlings. It is difficult to restore the relationship between a community and a plant resource, even one that has developed over centuries, once the resource has been depleted and the environment has changed, and people have largely forgotten the benefits of sustainable plant use.

THREE

Camu-camu

Fruits, Floods, and Vitamin C

Sahua cocha and IIAP field station at Jenaro Herrera (S4°54′21″, W73°40′07″), fourteen hours up the Ucayali River from Iquitos, Peru, 1984–1987, 2011

I was on the lookout for interesting fruits from the moment I arrived at Iquitos, and the first one that caught my eye was a small red-and-green marble to golf ball–sized fruit called *camu-camu* (*Myrciaria dubia*).[1] I saw it everywhere. Street vendors pushed carts loaded with baskets of camu-camu around town, women in the central market piled up plastic bags filled with the fruit for sale, delicious camu-camu juice drinks were available all over the place, and camu-camu sherbet was one of the first flavors listed on the signs outside the local ice-cream parlors. This was clearly one of the most popular native fruits in town. My first question was, I wonder where all these fruits come from?

(*Opposite*) Harvesting the fruits of *camu-camu* (*Myrciaria dubia*) in Sahua cocha, Department of Loreto, Peru. The tops of the flooded camu-camu plants are visible in the background; the floodwaters were about two meters deep at the time of the photograph.

I went around town and asked the street vendors, juice makers, market women, and ice-cream scoopers where they got their camu-camu fruit. What I really wanted to know was whether it was cultivated locally or "wild harvested" from the forest; I was hoping the latter. My informants all told me that they bought the fruit from sellers at the port,[2] and that it came from someplace "upriver." I started hanging around the port, and soon saw several *chaucheros* (dockworkers) climbing up the slippery bank from the river with huge bamboo baskets on their backs filled with camu-camu fruit.[3] Once they had reached the top and put down their baskets, I began interviewing them. All the chaucheros told me the same thing: "The best camu-camu comes from the oxbow lakes outside Jenaro Herrera, about fourteen hours from Iquitos up the Ucayali River."[4]

I had come to Peru to participate in a three-year study of the ecology and management of native fruit trees. As luck would have it, the local research institute (Instituto de Investigaciones de la Amazonía Peruana) with which I was collaborating had a small field station outside Jenaro Herrera. And it did take fourteen hours—and an overnight ride on a riverboat packed with swinging hammocks, motorcycles, innumerable sacks filled with rice, tubers, and electronic equipment, lots of children and babies, and a few water buffaloes—to get there. The station was located five kilometers from the town of Jenaro Herrera up a dirt road. About a dozen of us, mostly students and a few researchers, lived at the station, and I had a comfortable, palm-thatched hut with bathroom, running water, and even electricity for two or three hours most evenings. I also had the use of a wooden longboat named *Myrciaria* that had been made by a local boat builder and an incredibly knowledgeable and pleasant field assistant named Umberto Pacaya, who was a skilled boat driver, knew where to find all of the native fruits, and never once got lost in the forest. It was a great place to work.

On my first field excursion, I went to explore the two big oxbow lakes, Supay cocha and Sahau cocha, outside Jenaro Herrera, where the camu-camu was growing.[5] The bank of each of the lakes had a 20–30-meter-wide strip packed with camu-camu shrubs. When the water level in the lake was up, it was hard to paddle through the canopy of the camu-camal (dense aggregations of camu-camu); when the water level was down, we had to crawl, climb, and pick our way through the tangle of stems with care. As soon as the water level in the lake had dropped enough to uncover the entire camu-camu population, we started laying out our inventory plots.

The plots were laid along a straight line with the edge of the lake on one side and the levee—where flooded forest was growing—on the other. We had to work fast because the water level in the lake was rising a little each day and the front of our plots was becoming submerged. We were able to finish ten plots before they were all flooded. We measured and tagged almost nine hundred camu-camu plants in the plots; over half of these were seedlings and saplings. To put this number in perspective, most tropical forest trees form natural populations with only a couple of reproductive adults per hectare; the camu-camu population I had put my plots in had an estimated thirty-eight hundred fruit-producing plants per hectare.

The camu-camu plants started flowering while we were still working on our plots. The species produces beautiful white blossoms with a pleasing fragrance and sweet nectar to attract bees for pollination, and bees constantly buzzed around us as we worked. As the water continued to rise, the trees started to form small fruits, and by the time the water level had reached a couple of meters, the crowns of the trees were completely covered with shiny red fruit. These either fell into the water, where they were subsequently eaten and dispersed by fish, or were collected by local villagers.[6] Within a week or so, all the trees were completely underwater.

Although camu-camu is a terrestrial plant, it spends six to seven months each year in this state. Its life cycle is tied to the rise and fall of the oxbow lake, and in the few months it is out of the water, seeds have to germinate, saplings have to grow, and adult trees have to produce flowers; these have to be pollinated, and a large quantity of fruit must then be nourished to maturity—all before the waters rise again. The full sun and rich alluvial soils undoubtedly help the species fulfill these biological necessities in such a short period of time. Few other woody species can tolerate the severe annual flooding, so the camu-camu essentially has these fertile sites all to itself—another major advantage.

One of the important goals of my research was to estimate how much fruit the camu-camu populations were producing. This turned out to be somewhat complicated. When measuring the production of fruit by a plant population, the researcher usually tries to quantify the *size-specific* production: the number of fruits produced by plants of differing size. With this result, the researcher will multiply the fruit-production values for each size class by the number of plants recorded in that class to estimate total fruit production by the population. It is thus important not to mix the fruiting branches from big trees with those from smaller individuals. When water levels are low and the trees are completely uncovered, it is easy to see which fruiting branches belong to each individual. But when the water level rises, and it becomes harder to see which trunk a given branch is connected to, researchers can often make mistakes. To overcome this difficulty, we decided to number and label the branches pertaining to each sample tree with flags, and then move all the flags up as the water level rose. In this way, even when we could not see the main trunk of the tree, we could distinguish and separate out all of its fruiting branches.

Umberto Pacaya and I flagged the fruiting branches on twenty-five camu-camu trees of differing sizes and untied and retied hundreds of branch flags as the water level in the lake rose. We continued doing this for two years; the results were impressive.[7] The natural populations of camu-camu growing at the lake produced between 1.2 million and 1.6 million fruits per hectare each year. Based on an average fruit weight of about eight grams, this represents an annual production of from 9.5 to 12.7 metric tons of fruit per hectare. The majority of these fruits were produced by individuals in the smaller size classes, those with basal diameters of two to four centimeters. These individuals were also the shortest trees in the population, and as a result they were out of the water for only three or four months each year.

The fruit of camu-camu is *extremely* acidic. I ate so many fruits as I bobbed around in a boat in the middle of an oxbow lake in Peruvian Amazonia, that my lips were covered with blisters. Curious about where the acidity came from, I collected some fruit and sent it to a laboratory in Lima for nutritional analysis. The results made it clear that the acidity was caused by the high concentration of ascorbic acid, vitamin C. The fruit contains 2,000–3,000 milligrams of ascorbic acid per 100 grams of pulp; eating three fruits is roughly equivalent to taking a 500-milligram tablet of vitamin C. Oranges, by contrast, which are touted for their high vitamin C content, contain only 30 milligrams of ascorbic acid per 100 grams of pulp—about a hundredth of that found in camu-camu.

The wild stands of camu-camu growing along the oxbow lakes in Peru are some of the highest-density aggregations of a single plant species found in the tropics. These stands are extremely productive, largely because they are naturally "fertilized" each year by the floodwaters of the Ucayali River. They produce a fruit that is not only tasty but also extremely rich in vitamin C, and local communities actively exploit them for the revenues that the sale of the fruit provides. Collectors shipped over 45 metric tons of camu-camu fruit from Jenaro Herrera to Iquitos in 1984, for example. These populations, in essence, are huge, organic factories of vitamin C. They cost nothing to maintain. And they will continue to produce for as long as the ecological requirements of the species and the habitat are maintained, the river continues to rise and fall in a somewhat predictable manner, and the villagers are careful to harvest a sustainable amount each year, leaving a percentage of the seeds on the site to facilitate the regeneration of the population.

Local collectors have been harvesting camu-camu fruit for commercial use from the oxbow lakes outside Jenaro Herrera for several decades. The species would seem to be more resistant to the effects of harvesting because of its high-density populations and abundant fruit production, but until recently nothing was known about the impact of fruit collection on the structure and function of wild camu-camu stands. Market demand for the fruit was still strong in Iquitos, and villagers continued to paddle out to the lake and fill their canoes with camu-camu fruit each year. What effect was this having on the plants?

To address this question, a graduate student of mine, Meredith Martin, traveled to Jenaro Herrera in 2011 to resample my plots at the oxbow lake, twenty-seven years after the plots were originally surveyed. The camu-camu population had been harvested for fruit every year since then, at increasing intensities and by more people as the market expanded. Martin found my field assistant, Umberto Pacaya, now a bit older, but still thrilled at the prospect of fieldwork and per diems, and he remembered exactly where the plots were. Together they laid out a new set of plots, as close as possible to the original location, and recounted and remeasured the camu-camu plants. What they found was that the density of camu-camu individuals on the site had decreased by about 75 percent.[8] While it might seem obvious that excessive harvesting was the cause of the decrease, another riparian shrub, *Eugenia inundata*, known locally as *fanache*,[9] found in both the original plots and the second inventory, also exhibited a significant decrease in stem density—and this species had not been harvested.

Something other than commercial fruit collection seemed to be affecting the regeneration of both shrub species. A look at the dynamics of the Ucayali River over the past two decades suggests a plausible explanation. Since the late 1980s, the frequency of extreme hydrological events in the Amazon Basin has increased significantly. Severe droughts and high flooding have been recorded several times, and the flood of 2009 was ranked as one of the highest and longest of the past 107 years. Although riparian shrubs are well adapted to the unpredictability of life along the floodplain, they need sufficient time out of the water to flower, fruit, and establish new seedlings. Commercial harvesting decreases the number of seeds available for germination and establishment; too much time underwater has an impact on both the reproduction and the growth dynamics of the population.

The changes noted in the density of the camu-camu population might also be linked to the inevitable successional development of the oxbow lake. The lakes at Jenaro Herrera remain connected to the Ucayali River through a tie channel, or *caño*. These channels transfer water and sediments during the flood cycle and act as the dominant mechanism of lake infilling. The lake is gradually filling up with sediment, the surrounding levees are slowly shifting inward, and tree species from the nearby flooded forest are moving in to colonize the camu-camu site.

There is no question that commercial fruit collection has had an impact on the regeneration and population structure of camu-camu. Other factors, however, are also at work. The oxbow lake is filling up, new species are moving in, and it is only a matter of time before the lake—and the camu-camu that grew there—will be gone. The extreme flooding that has occurred in recent years will undoubtedly lead to the formation of new oxbow lakes with new riparian habitats to colonize. If they have a source of seeds and the fish to disperse them, camu-camu will start a new population. The fruit, the collectors, the oxbow lake, and the river are all part of a particular context in the floodplain of Peruvian Amazonia. The work with camu-camu at Jenaro Herrera is a useful reminder that contexts change.

FOUR

Fruits from the Amazon Floodplain

Iquitos, Peru (S3°44′56″, W73°15′13″) and environs, 1984–1987

During my three years in Peruvian Amazonia, I spent a lot of time in the Belen market of Iquitos—a twenty-block, byzantine sprawl of stores, stalls, houses, and canoes, reached by steep concrete stairs. During certain times of the year, much of the market is underwater, but residents of this part of Iquitos appeared to have developed two strategies for dealing with the annual flooding of the Ucayali River. They either built their houses on pontoons that would gradually start to float as the floodwaters rose, or they lived in two-story houses and moved upstairs for the duration. The latter required more work because they had to dig the mud out of the first floor every year after the floodwaters receded. Yet inhabitants of both types of houses, surprisingly, would keep their electrical lines con-

(*Opposite*) Fruits of the *ñejilla* palm (*Bactris brongniartii*) for sale in the Belen market of Iquitos, Peru.

nected, live, and dangling during the flooding. They might be floating on the tide or have a basement full of water and mud—but they still had electricity to run their stereos.

I saw a number of unexpected sights in this market. I saw a man carrying a catfish that was bigger than me in a basket on his back. I saw several species of monkeys being grilled over a brazier; bundles of hand-rolled cigarettes, known as *mapachos*, used for smoking, getting a botfly out of an arm, or casting a spell; grimy glass bottles filled with cane alcohol infused with strips of bark for improving one's sex life or, conversely, for use as a contraceptive; and stacks of T-shirts made in China printed with cryptic slogans in questionable English. But the main item of interest in the market was the dazzling diversity of native fruits. There were dozens of different types of fruits, and the variety changed from month to month. My colleague in Iquitos, Christine Padoch, who was investigating the production and marketing of fruits while I focused on wild fruit trees, counted over a hundred different species of native fruits being offered for sale in the market. The majority of these fruits were wild harvested.

The fruits of *sacha mangua*, or "false mango" (*Grias peruviana*)[1] have a large seed, carrot-colored flesh, and a fuzzy brown covering; they are the size and shape of a large potato. They are eaten peeled, with slices of the stiff, orange flesh separated from the seed, either raw with *fariña*—granules of toasted manioc (*Manihot esculenta*)—or grilled. An oil can also be extracted by boiling the fruit; the seed, which contains saponins (foamy glycosides), is used as soap in many rural areas. Nutritional analyses have shown that the fruits are rich in vitamin A.

Sacha mangua trees grow in the understory of seasonally flooded forests in Peruvian Amazonia, where they can form dense aggregations containing two hundred adult trees per hectare. The tree has a distinctive growth form, and it is easy to recognize in the forest.

Adult trees are rarely taller than 20 meters, and the slender, unbranched stems terminate in a cluster of one-meter-long spatula-shaped leaves. The flowers and fruits grow directly out of the trunk —a bit like a tree designed by Dr. Seuss. Given the ease with which the tree, its seedlings, and its saplings can be recognized, the use and value of its fruits, and the propensity of the species to form dense stands, I selected sacha mangua as one of my study species. I laid out eight plots in a tract of flooded forest located about a kilometer up a blackwater tributary of the Ucayali River and started counting and labeling trees.[2] Quantifying the production of fruits was relatively straightforward. I selected fifteen adult trees, visited them every week, and counted the number of new fruits along the stem of each tree. I marked the new fruits with paint each week to avoid duplication in subsequent counts.

I noticed during my weekly visits to the sample trees that several of the immature sacha mangua fruits had scratches on them. Each week there would be one more scratch—I counted them—on the fruits. This continued for several weeks until the fruits were mature. At that time, when I visited the tree all that was left was the stalk of the fruit, most of the brown covering, and a little orange pulp. The entire seed was gone. My field assistant Umberto finally explained what was going on. Squirrels were also visiting my sample trees each week and making small scratches to see if the fruits were ripe yet, like savvy shoppers testing cantaloupes in the produce aisle. Once they saw that the fruits were mature or, more important, that the seeds were fully formed, they gnawed into them and made off with the seeds.

I continued monitoring my plots for about a year and a half and compiled a complete set of demographic data for the species, consisting of the number of individuals in different size classes, how many died and how fast the rest grew into the next size class, and the size-specific fruit production. I put all these data into a computer model to assess the stability of the population and to perform a series of simulations to estimate how many fruits could be harvested without affecting the long-term stability of the population.[3]

The sacha mangua population contained over five hundred individuals with a plethora of seedlings and saplings; the species was actively regenerating itself on my site. The adult trees in the population flowered and fruited almost constantly throughout the year, except during the flood peak and for a couple of months thereafter. All together, they produced more than eight thousand fruits per hectare—minus the ones the squirrels made off with. The total annual production of fruits was in excess of 2.5 metric tons per year.

My computer simulations revealed that the sacha mangua population was close to stability, in fact was gradually increasing in size each year. The simulations also showed that collectors could harvest about 80 percent of the fruit produced each year, roughly sixty-five hundred fruits, with little impact on the long-term stability of the population. Based on this finding, I became curious to see what level of harvest *would* overexploit the population. How many fruits would I have to collect to drive the population to extinction? A morbid question, I know, but I was only doing a computer simulation. In my model I simulated a harvest of 95 percent of the fruits each year, ran the program to calculate the resultant size of the population, and continued doing this for eighty years. The results from this rather tedious exercise were not what I had expected.

The population was able to maintain its stability, with birthrates and death rates in the population balanced, on the site for almost thirty years. Levels of seedling establishment were gradually dropping during this period, but dead trees in the larger size classes were continually being replaced by smaller trees. This changed as I continued the simulation. Adult trees started to die, the rate of seedling establishment plummeted, and the seedling and sapling classes began to empty out completely. There were still large adult trees that were producing harvestable quantities of fruit, but the population was recruiting no new individuals each year. This meant that a fruit collector who was only looking at the adult trees might conclude that everything was fine. But a plant ecologist, looking around in vain for seedlings, saplings, and young sacha mangua poles, would immediately appreciate that a crisis was at hand. After about fifty years, even a fruit collector would notice that something was wrong because there would not be much left to harvest. After eighty years, the last sacha mangua tree disappeared from the site—in my computer simulation.

Harvesting too much from a plant population can drive it to extinction. But unless the population is monitored carefully, harvesters might not notice the impact their fruit collecting was having for forty or fifty years. I have a feeling that we are at this stage with a number of the more valuable wild harvested fruits in Amazonia, such as the *aguaje* palm.

The fruits of the aguaje palm (*Mauritia flexuosa*)[4] are highly esteemed in Iquitos. They are eaten raw or processed into a sweet paste for beverages and ice cream, and the pulp and seed kernel yield an edible oil of good quality. Aguaje fruits have a large seed, a thin layer of tasty orange pulp, and a tight, protective covering of small red scales. To get to the sweet part of the fruit, you have to nibble off the scales one by one, spit them out, and then scrape the fleshy orange pulp off the seed with your teeth. It takes some practice to learn how to do this with grace, and eating an aguaje fruit is not something one does in a hurry. They say in Iquitos that the easiest way to get a girl to chat with you for a few minutes is to buy her an aguaje fruit.

Aguaje is a massive palm with large fan-shaped leaves. It is easy to identify in the forest, especially since it frequently occurs in locations that are sparsely populated with other trees. Aguaje is one of the few species that can tolerate the waterlogged conditions in the swamps and permanently flooded backwater forests of Amazonia, and over a million hectares of *aguajales* (dense aggregations of aguaje palms) are found in Peru alone. These aggregations can contain 150 adults and 400–500 aguaje seedlings, saplings, and prereproductive palms per hectare. Not only are wild populations of aguaje exceedingly dense, the large number of individuals in the smaller size classes suggests that the species is regenerating itself in the swamps at a rate sufficient to maintain the population at its current density. The adults in these populations produce a staggering amount of fruit—over six metric tons per hectare in some of the populations that have been studied in Peru.[5] The aguaje, then, is an extremely well adapted palm that produces large quantities of valuable market fruit and forms high-density aggregations in habitats that are not suited to other forms of land use.

Fruit collectors in Peru intensively exploit the local aguaje forests. The average demand for aguaje fruit in Iquitos has been estimated at about 15 metric tons per day, and the majority of the fruit is wild harvested.[6] While the intensity of exploitation is not in itself a problem, a key aspect of the biology of the palm coupled with the destructive, short-sighted method collectors currently use to harvest the fruits has created a relationship between people and plant that is decidedly dysfunctional.

Aguaje palms have separate male and female trees. Only about half the trees in the population will produce fruit; the remaining trees produce the pollen required to complete the reproductive process.[7] Fruit collectors have developed several methods of harvesting the fruits from tropical trees. They pick the fruit up after it falls from the tree, knock it out of the tree with a long pole, climb the tree to collect it (native Amazonians have devised a number of ingenious ways to scale fruit trees using rope belts and slings for their feet), or, as seems to happen with increasing frequency these days, cut the entire tree down. The latter method, unfortunately, is especially common with palms that have large, heavy bunches of fruit and thick trunks, and are extremely tall—like the aguaje palm. Most of the aguaje fruits offered for sale in the market and on the streets of Iquitos were harvested by cutting the palm down. But only the female palms are felled. The aguajales and palm swamps that were once extensive in the vicinity of Iquitos—the swaths of forest that look so photogenic from an airplane—are actually barren stands of male aguaje trees. They contain no fruits, no seeds, and no seedlings, and, before long, they won't contain any aguaje palms either.

The growth form of a palm can have a major influence on the pattern, intensity, and ultimate sustainability of its use by local communities. A comparison of two closely related palm species, *açaí* and *huasai*, offers a useful example. Açaí (*Euterpe oleracea*)[8] is a slender, multistemmed palm that forms dense aggregations in the Amazon estuary and along whitewater rivers in eastern Brazil. The species is an important source of palm hearts, and a beverage made from the fruit pulp is a stable component of the regional diet. The huasai palm (*Euterpe precatoria*)[9] grows in similar environments in western Amazonia, also occurs in dense, natural population, and also produces palm hearts and edible fruits. The salient difference between the two species is that huasai is a single-stem palm.

Palms, unlike broadleaf trees, have only one primary meristem, and this important collection of cells is located at the apex of the plant where it elongates the stem and makes new leaves.[10] In addition, multistemmed palms have buds at the base of the stem that can be activated to produce additional shoots—which have only a single primary meristem. Single-stemmed palms do not have these buds. When we order palm hearts in our salad we are getting the primary meristem of a palm, and the extraction of palm hearts from a single-stemmed palm kills the plant. Extracting palm hearts from a multistemmed palm kills the shoot on which the heart grows, but triggers the basal buds that will eventually produce new shoots.

Palm hearts can be extracted from several different species, but rarely will palm hearts from huasai appear on the menu. Not enough of these palms grow at a reasonable distance from Iquitos to supply the market. They did at one time, but then a palm heart canning factory was opened in Iquitos. The factory was eventually forced to close for lack of raw material after the huasai palms were completely harvested from local forests. Commercial exploitation of wild populations of a single-stemmed palm for palm hearts is risky business.

Forest farmers in Brazil, on the other hand, have developed sophisticated systems for managing wild populations of açaí palms for both palm hearts and fruits. Closely controlled palm heart extraction is used to stimulate the production of new fruit-bearing shoots that are harvested every year. There are extensive stands of açaí palm throughout the Amazon estuary, and more are being created through enrichment planting and forest refinement to supply the growing market for the fruit. Forest habitats are being sustainably managed and conserved, and local communities are generating additional income for themselves. The multistemmed growth habit of the açaí palm—its ability to re-sprout after cutting—is largely responsible for this.

As noted earlier, the best-known, most widely cited characteristic of Amazonian forests is the large number of different tree species that grow in them. Small tracts of forest may contain several hundred tree species per hectare, and new taxa are being discovered every year. An unavoidable correlate with high species diversity, however, is that the trees of a given species are usually scattered throughout the forest at relatively low densities. Many common trees occur at densities of only one or two individuals per hectare, and some of the rarer species occur at even lower densities.

The scattered distribution and low abundance of many Amazonian tree species represent a dilemma for local collectors. Harvesting takes considerably more time and effort when the trees are spaced far apart, and yields per unit area are likely to be low. Tree populations with only a few adult trees also have a limited capacity for regeneration, are extremely sensitive to the effects of destructive harvesting, and can easily be overexploited. Several commercial forest resources have already been depleted through overexploitation, and many of the more valuable tree species are currently grown in plantations or small-scale agroforestry systems, not wild harvested, testimony to the difficulties of extracting commercial quantities of useful products from species-rich forests.

All the Amazonian fruit species mentioned so far—camu-camu, sacha mangua, aguaje, huasai, and açaí—form dense, natural populations in habitats where seasonal flooding or permanent swamp conditions preclude the formation of more species-rich mixtures. The oligarchic forests—forests dominated by a small number of species—formed by these species are extremely dense, highly productive, and contain more juveniles than adult trees, suggesting that the populations are actively regenerating themselves.[11] In terms of both density and yield, the oligarchic forests of native fruits in Amazonia rival many of the commercial fruit orchards that have been established in the tropics.

It is puzzling, therefore, that the extractive reserves and community forestry initiatives in the region have invariably been based in species-rich forests. It would seem more logical to focus attention on forest ecosystems that have the greatest potential to sustain a program of commercial resource use. In terms of density of useful products, productivity, familiarity to local communities, and relative ease of management, oligarchic forests are much better suited to this form of land use than other types of plant community in Amazonia.

A number of promising native fruits in flooded forests do not form dense populations, however. *Uvos* (*Spondias mombin*),[12] for example, usually occurs at densities of five to ten trees per hectare, and finding them takes some work. But it is worth the effort. Uvos trees produce large numbers of bright orange, grape-sized fruits that have an astringent and pleasingly tart flavor. The seeds are fibrous and the fruits float. The trees flower when water levels are low and they drop their fruit into the water during the flooding season. Paddling through a flooded forest and coming upon thousands of florescent orange uvos fruits bobbing up and down in the water, then reaching out and scooping up handfuls to eat is one of the joys of doing ecological research in lowland Amazonia.

Local people use almost every part of the uvos tree. The fresh fruits are eaten raw or made into juice, ice cream, and jellies; the wood is used in light construction; and a tea made from the leaves and bark is considered medicinal. An alcoholic infusion of the bark is said to be an effective contraceptive. I remember that one of the *materos* (woodsmen) who worked with me would strip off a piece of bark every time we walked by an uvos tree and extol the virtues of the species as a means of birth control.[13] I would always nod my head, but I maintained a certain degree of skepticism, because I frequently saw his wife and ten children in the village.

The *huito* tree (*Genipa americana*)[14] also produces a fruit that is collected and sold in local markets. While the fruits are edible and are used to make juice and ice cream, a somewhat more unusual use is as a body paint. When the juicy liquid from the fruit oxidizes, it stains the skin dark purple or black. Indigenous Amazonian men use the juice to paint their faces before hunting, going into battle, or visiting a girl friend. The dye obtained from huito is essentially permanent, which explains the large number of tourists with black lines on their faces getting on the airplane in Iquitos to go home after a visit to a "traditional" Amazonian village.

In spite of the relatively low densities of uvos and huito trees in flooded forests, both species appear to be actively regenerating and maintaining their populations. The high demand for uvos ice cream and juice in Iquitos drives collectors to harvest large quantities of the fruit, but apparently a sufficient number of seeds escape collection and germinate to produce a new crop of seedlings each year when the river level drops. The demand for huito fruits is much lower. Only a few vendors in the market sell huito, and the juicy exudate from a single fruit is sufficient to paint several faces. Many of the fruits produced by this tree species remain in the forest.

I became friends with a well-known botanist from Missouri, Al Gentry, who would occasionally visit Iquitos, and during one of our dinners together we discussed the idea of trying to put an economic value on all the non-timber resources obtainable from a tropical forest. He had just finished a detailed inventory of a tract of forest near Iquitos, and I had been collecting data on fruit production by forest trees and had been following the market price of different forest products for several years. I still remembered a little from the resource economics class that I took in forestry school, and I knew that if we could quantify the number of fruit-producing trees in the forest and estimate how many fruits each one produced in a year, then put these two numbers together and multiply the result by the market price for each resource, we could come up with an estimate of the value of one year of fruit production in the forest. If we did this for a number of years and brought the combined value of all future harvests back to the present using a discount rate, we could also calculate the net present value, or NPV, of the forest. If the same measurements were made for alternative land use scenarios such as logging or clearing the forest to make a cattle pasture, a comparison of the NPVs should show which form of land use represented the best investment.

We had both seen the bustling commerce in the local market and knew that the collection of forest fruits was an important part of the livelihoods of local communities. We also knew that a forest exploited for non-timber resources seemed to be in considerably better shape than one that had been logged or cleared to make a plantation or a pasture. We enlisted the help of an economist to help us with the calculations, wrote a small opinion piece, and submitted the manuscript to a prestigious scientific journal. It was accepted and published in June 1989.[15]

What we found out in our analyses was that the net present value of the tropical forest outside Iquitos, harvested annually for fruits and latex, was a little over $6,000. Selectively logging the timber on a twenty-year rotation yielded a net present value of $490; clearing the forest to make a plantation of fast-growing timber trees or create a pasture both generated net present values of about $3,000—half of the NPV calculated for the sustainable harvest of non-timber forest resources from the forest. We pointed out that harvesting for fruit and latex appeared to be the best way to use a tropical forest. The harvest of fruit and latex yielded higher net revenues than timber, and the two resources could be harvested with considerably less damage to the forest. We closed the piece by suggesting that comparative economics might provide the most convincing justification for the conservation and sustainable use of tropical forests.

Soon after our commentary was published, prominent pieces about our findings appeared in the *New York Times* and the *Washington Post*.[16] I was interviewed on National Public Radio, and copies of the article, I was told, were distributed to all the delegates at the United Nations. I got a lot of telephone calls. And, of course, we also received the requisite criticisms of the piece because our study forest was located so close to Iquitos, and the discount rate could have been different: if the collectors had harvested all the fruit from all the forests the market value of these products would drop to zero, and different forests with different floristic compositions would surely exhibit a different net present value. I do not dispute any of these criticisms, and acknowledge that we could have done things differently. That said, the collection of forest fruit *does* generate a significant amount of income to local communities in Peruvian Amazonia. Non-timber forest products, like fruits, *can* be harvested from tropical forests with less ecological impact than harvesting timber, and comparative economics *always* provides the most convincing explanation for the land-use decisions that human beings make. That a tropical ecologist and a botanist were the first people to bring up these issues paints a pretty clear picture of the state of non-timber resource and sustainable forest use thirty years ago. A lot of people still think I'm an economist.

FIVE

Forest Fruits of Borneo

Pontianak (S0°1′21″; E109°19′49″), West Kalimantan, Indonesia, and assorted locales in Sambas, Sanggau, and Kapuas Hulu Regencies, 1989–1994

The winged seeds produced by several species of trees in Borneo contain an edible oil whose physical and chemical properties are remarkably similar to cocoa butter. The seeds are known locally as *illipe* nuts, and large quantities are collected and sold to be used in the manufacture of chocolate, soap, candles, and cosmetics. The triglyceride fractions in illipe nut oil occur in similar proportions to those found in cocoa butter, and the oil can be blended with chocolate without altering the texture, gloss, or taste of the original confection. The higher melting point of illipe nut oil makes it especially useful as a chocolate hardener or to provide added body

(*Opposite*) Dayak children in West Kalimantan, Indonesia, holding the winged fruits of *tengkawang* (*Shorea macrophylla*).

to face creams, so that the chocolate, as they say, "melts in your mouth, not in your hand," or the cream stays on your face during a hot day at the beach.

Illipe nuts have five wings and come in a variety of sizes and shapes. Some of the smaller seeds are the size of a marble, and the long wings twirl the falling seed like a helicopter, helping to disperse it. The larger-seeded varieties are the size and weight of a golf ball, and the short wings on these seeds do little to prevent them from plummeting straight to the ground directly under the crown of the parent tree. Regardless of the size of the seed produced, however, all species of illipe nut trees flower and fruit synchronously at unpredictable intervals of from two to ten years. This reproductive behavior, which is known as mast fruiting, is thought to limit the abundance of frugivores and seed predators in the forest by concentrating their food supply into a single brief period every few years when they are confronted with more food than they can possibly eat. Some seeds will inevitably be left uneaten on the forest floor to germinate and recruit new seedlings into the population.

As can be appreciated, this unpredictable reproductive strategy creates serious supply problems in the illipe nut market. Neither the buyers nor the sellers know exactly when the next fruit crop will be produced, and neither the dried seeds nor the oil can be stored for significant periods of time.

I had been thinking a lot about the illipe nut market when I arrived at West Kalimantan—Borneo, the third largest island in the world, is divided into three countries: Malaysia (Sabah and Sarawak) and Brunei in the north, and Indonesia (West Kalimantan, East Kalimantan, Central Kalimantan, South Kalimantan, and North Kalimantan) to the south—in 1989 to continue my investigations on the ecology and management of tropical forest fruit trees. It seemed to me that finding an illipe nut tree that fruited every year would solve both the buyers' and the sellers' problems.

On my first trip to the field, I went *by cab* from Pontianak to the Raya-Pasi Nature Reserve in the Sambas district, a roughly two-and-a-half-hour ride. We stopped in Bagak Sahwa, a small village near the reserve, to pick up the park warden. As we were walking up the mountain into the reserve, I casually asked the warden whether he knew of any annually fruiting *tengkawang*, or illipe nut, trees that grew in the reserve. He said that he did, and took me to see several dozen large illipe nut trees. The forest floor was carpeted with the seedlings and saplings of this species, *Shorea atrinervosa*,[1] which I found surprising, because the previous year had not been a mast year and none of the other illipe nut species in West Kalimantan had produced any fruit.

I spent the next three years studying a marked population of these illipe nut trees in the Raya-Pasi Nature Reserve to determine whether the warden was right about the annual fruiting behavior of these trees.[2] My study plot contained over 180 species of trees.[3] Included among these were 25 species that produce edible fruits or nuts, 35 timber species, 5 species that produce a valuable oleoresin,[4] 3 species whose leaves and bark are used medicinally, and a species whose fruits are used as a fish poison. There were 19 adult trees, several thousand pole-sized individuals, 7,000 saplings, and 50,000 seedlings of that illipe nut species in my one-hectare study plot—the population was experiencing a huge surge of growth. This made sense. If the illipe nut trees were producing fruit every year while the associated tree species were mast fruiting, the former were producing many more seeds and establishing orders of magnitude more seedlings each year. Additionally, I never saw hordes of wild pigs ravaging the fruits on the ground, so perhaps the selective advantage afforded by mast fruiting—to satiate seed predators—no longer operated on the site.

I obsessively followed the reproductive behavior of the adult illipe nut trees in the population for two years. All together, they produced over 160,000 illipe nuts—equivalent to a total production of about one metric ton of fresh fruit—per hectare. *And they did this both years.* The warden was right.

In addition to working with the warden, I did much of my fieldwork at Raya-Pasi with an older Dayak (a collective term for the indigenous people of Borneo) man who knew all the local trees. He would show up for work in shorts, a T-shirt, and flip-flops, with a dingy white towel wrapped around his head. He carried a razor-sharp *parang* (bush knife) in his hand and always had a crumpled plastic bag in his pocket with a box of matches and a pack of clove cigarettes (*kreteks*).[5]

He was extremely hard working and never stopped telling stories, such as the time when he was a boy and the Japanese army marched into Bagak Sahwa; everyone except the village leaders ran into the forest. He told me about the first time he saw a car. He described how the local people would cut off the buttresses of certain trees to make pans for mining gold, and why everyone in the village had left the new water hydrants put in by the development project running all the time until the pump burned out—because water is supposed to flow. And he related wonderfully detailed, ethnobotanical anecdotes about every one of the tree species we measured and tagged.

The best times came when the wind would pick up and the rain would start pouring down. We would squat at the base of the nearest buttressed tree, and I would dig out the rain poncho from my daypack and try to cover us with it. After a few minutes, once it had become really hot and stuffy under the poncho, he would pull out his plastic bag and light up a kretek. And we would soon be sitting in a dense cloud of clove smoke. Although I could no longer see his face, I could still hear him telling me his stories.

The city of Pontianak has eight different produce markets, and all of them sell native fruits. The fruit offerings vary throughout the year, but if you visit the markets every couple of weeks for several years, you will encounter almost a hundred different species. There are breadfruits (*Artocarpus heterophyllus*), jackfruits (*A. altilis*), and *cempadak* (*A. integer*), closely related fruits in the Moraceae family; numerous varieties of mangos (*Mangifera* spp.); several different types of rambutan (*Nephelium* spp.); stinky beans (*Parkia speciosa*) that make your urine smell funny; durians (*Durio zibethinus*), mangosteens (*Garcinia mangostana*), and crunchy fruits from spiny palms (*Salacca zalacca*); and always a large assortment of fruits (*Baccaurea* spp.) that when peeled reveal a neatly packed cluster of seeds, each surrounded by a layer of delicious, fleshy pulp.[6] Pop the whole fruit in your mouth, suck on the pulp, and spit the seeds out one by one.

Unlike in the central market in Iquitos, where a similar diversity of native fruits, mainly wild harvested, was observed, all the fruits offered for sale in Pontianak were cultivated to some extent. Even the commercial fruit trees in the forest were planted. Or at least, someone spit out the seed there.

Durian is unquestionably the premier market fruit of Southeast Asia. The fruit is the size of a large football and has a thick, woody husk covered with sharp thorns. It has several sections, each containing a row of seeds covered with thick, creamy pulp. The pulp, the edible part, emits a distinctive odor that people find either extremely agreeable or completely disgusting. I personally love the fruit, but it is definitely an acquired taste. My favorite description of the flavor was a colleague's "it tastes like French custard that has been dragged through a toilet."[7]

Durian can be counted on to be a conversation starter in Pontianak. In the weeks leading up to the fruiting season, everyone speculates about the harvest. Will the fruits be large, meaty, and aromatic? Will the price be as good as last year? Which villages will bring the best fruits to market? Once the durians start showing up in town, the conversation turns to where to find the tastiest and best-priced varieties. The answer can change from week to week during the fruiting season, so people talk about it *all the time*. Once the season has ended, people reflect on the quality and quantity of the fruit from the past season, sharing anecdotes such as the way a seller was tricked into accepting a lower price by a long-time resident, or how a local durian connoisseur went through the entire season and never selected a bad fruit. And this conversation continues until the next fruiting season.

During durian season, the sellers start piling their fruits onto plastic tarps along the streets in the late afternoon to keep them from the hot sun, which would cause them to dry out or even to split open. Some areas of town are more desirable for market stalls than others, busy streets with lots of traffic, for example, and the more senior and successful durian sellers always claim these spots.

The preferred evening activity in Pontianak during durian season is to shop for fruits. After dinner, the entire family gets in the car or piles on the motorcycle or flags down a *becak* (three-wheeled cycle rickshaw) and heads downtown.[8] They slowly cruise around looking at the different stalls until they see a pile of fruits they like. The mother and father will talk to the vendors, whom they may know from previous seasons, and pick through the pile, hefting and smelling favored fruits. Animated discussions will ensue over each fruit; a lot of head nodding goes on. Once they have selected their fruits, they negotiate a price with the vendor, pile the fruits and the family into the car, the motorcycle, or the becak, drive home—and eat them *all.* At the end of the evening all that remains is a pile of thorny husks and seeds and the lingering odor of durian. The same scenario will be repeated the following night. And the night after that and the night after that until the durian season ends.

In an effort to determine how many durian fruits were being sold in Pontianak, in 1991 we toured the city every evening by motorcycle, stopping occasionally to talk to vendors and count fruits. The durian season lasted forty-five days that year, and the availability of fruits followed an expected pattern. Early and late in the season, several thousands fruits were for sale throughout the city. During the peak season, we counted over sixty thousand, and mountains of durians dotted the town. The selling price of the fruit, however, stayed relatively constant. The first and last fruits of the season cost only about twenty cents more than those available during peak fruit production. (Data show that similar patterns of inelastic demand are characteristic of the markets for illegal drugs and gasoline.) Pontianak produces an abundance of native fruits, but durian remains the number one choice of the locals.

Durians produce large, creamy-white flowers with abundant nectar and a distinctive, buttery smell. The flowers open at night and are pollinated exclusively by a single species of frugivorous cave bat (*Eonycteris spelaea*).[9] There are two fruiting seasons for durian in West Kalimantan, the main reproductive event and a smaller episode in which only a few trees bear fruit; each of these lasts approximately a month and a half. When the durian trees are flowering, they offer the bats an ample supply of nectar and pollen. But what do the bats live on for the rest of the year?

Apparently, during most of the year, the bats forage on a local mangrove species, *Sonneratia alba*,[10] that grows along the coast. These trees produce a small quantity of fragrant, night-blooming flowers more or less continually throughout the year. The bats fly to the coast each night, covering distances of up to forty kilometers, to feed on the mangrove flowers. On their way, they'll also check out any durian trees they pass for flowers, almost as a dietary afterthought. To put this in perspective, the exclusive pollinator of the king of market fruits in Southeast Asia, the source of a multimillion-dollar fruit trade, provides this service as a part-time job, a temporary, and somewhat unpredictable, supplement to its usual foraging behavior.

Things are changing fast for the cave bats, however, because the mangroves in West Kalimantan are being cleared to make room for coastal development. As the distribution and abundance of their main food source decreases, the number of bats flying though the forests of Borneo each night will also decrease. Fewer bats will lead to less pollination, and less pollination will lead to fewer fruits. Development activities far from where the fruits are growing can have a drastic impact on the productivity of durian. The farmers have no control over this—and might not even see it coming.

Madurese farmers from the village of Punggur, southwest of Pontianak, grow a delicious market fruit known as *langsat* (*Lansium domesticum*)[11] in the extensive peat swamps that surround their village. The small yellow fruit is produced in grapelike clusters; the edible portion is the translucent, whitish pulp that surrounds each seed.[12] The pulp tastes like a cross between a grape and a grapefruit. Savvy buyers in Pontianak specifically ask for langsat grown in Punggur, and the village makes a lot of money from the sale of this species.

The problem for the farmers in Punggur is that peat soils are too acidic for fruit trees. Peat is partially decomposed organic material that builds up in places where the rate of litter deposition exceeds the rate at which the material is broken down. This is the case in peat swamps because they have a low pH and reduced oxygen levels, both of which inhibit microbial activity and the rate of decomposition. In order to plant fruit trees, farmers need to go through a laborious process of hand digging a network of drainage channels through the peat.[13] Water from the spongy peat drains into the canals, and the peat slowly subsides. The farmers repeatedly burn off the top layer of dry peat, and clean out and deepen the canals, as little by little the peat continues to dry out and flatten. The objective of this backbreaking work is to eventually get rid of the peat and expose the fertile alluvial soil underneath. After ten years of this digging and burning, the peat has usually subsided enough for the farmers to plant rice and coconuts.[14] The peat continues to go down through successive rice harvests, and after five more years, the farmers can plant langsat trees. Durians, rambutans, bananas, several species of citrus, and a variety of other native fruits are also introduced into what has now become a species-rich, market-oriented, agroforestry system. Fifteen years later—a total of thirty years after clearing the forest and digging the canals—the farmers can start harvesting their fruit.

I don't know exactly how long it takes for peat to accumulate—peat accretion is reported to be on the order of one to two *millimeters* a year—but suffice it to say that it takes a long time to make a meter of peat. One of the farmers at Punggur recalled having to work through ten meters of peat before he could plant his langsat trees. Burning and digging, burning and digging, he finally reached the point where he could plant rice, and before much longer he caught a glimpse of mineral soil. Much to his surprise, after the last layer of peat was cleared away, he found fragments of tools and earthenware. Another group of farmers had been living in the same place, thousands of years ago, and had probably planted their own fruit trees—without having to deal with the peat.

Today's farmers' ingenious solution to their fruit-growing problems brings both advantages and disadvantages. Draining and burning several meters of organic peat soil releases an enormous amount of carbon into the atmosphere. From a global-warming perspective, this is not an advisable form of land use. But at the same time, the farmers in Punggur have created a completely new plant community in an extremely restrictive environment that contains multiple tree species and generates a continual source of revenue. These habitats will undoubtedly stay forested and productive—and thus growing and fixing carbon—for many generations. As long as the native fruits of Borneo continue to command a good price in local and international markets, the farmers in Punggur are likely to care for and harvest their langsat trees sustainably.

SIX

Homemade Dayak Forests

Various Dayak villages in Sambas, Sanggau, and Kapaus Hulu Regencies in West Kalimantan, Indonesia, 1989–1994

During my early days in West Kalimantan, I was always on the lookout for a good piece of forest to study. Driving up the coast from Pontianak I saw little intact vegetation, but soon after turning inland at the bustling market in Sungei Penyuh, I noticed several steep hills along the road that were densely covered with forest. Too small, too close to town, and too steep for me to climb every few weeks to use as a study site, they were nonetheless forested and definitely worth a look—at some point.

A few weeks later, in the early morning when the canopy trees were just emerging from the fog, I pulled off the road in front of a

(*Opposite*) Dayak man walking through his *tembawang*, or managed forest orchard, in West Kalimantan, Indonesia. Note the tapping scars on the rubber (*Hevea brasiliensis*) tree to the left, and the squirrel tail poking out of the woven palm-leaf basket.

small farmhouse and decided to go have a look. An older Dayak man came out of the house in a sarong, bearing a parang (bush knife) and a smile. We struck up a conversation, and I soon conveyed in my halting Bahasa that I was from the United States, that, yes, I was already married, and that I wanted to explore the forest on the hill behind his house.[1] As far as I could tell, he said something about his community controlling access to the forest, but that it was all right with him if I wanted to look around; he would show me the way. After I solicited his help in identifying the trees we might encounter, we hurried off to the forest up a rocky, steep, and virtually imperceptible trail.

I felt a notable drop in temperature when I moved under the canopy. The forest was relatively open, yet multistoried, with numerous palms and climbers. Many of the canopy trees were heavily buttressed and over a meter in diameter; I was immediately smitten. After noticing a few durian husks on the ground, I asked about some of the trees around us. My host knew them all, and he also knew when they produced fruit, how many baskets of fruit they yielded, what the market price for the fruit was last year, and roughly how old the trees were. *And then he told me the name of the person who had planted each tree.*

My field notes show that we walked through about two hectares of hill forest that day. I was shown five species of mango trees, seven species of breadfruit, six species of rambutan, eleven species of rattan, and three beehives. I counted forty-six large durian trees, sixteen sugar palms (*Arenga pinnata*),[2] and several dozen canopy trees that produce milky latex or other exudates of value. While I sat down to catch my breath, I reflected that local villagers were managing thirty to forty species of trees and an inestimable number of palms, climbers, shrubs, and herbaceous plants. To my untrained eye, I appeared to be in a beautiful, relatively undisturbed, piece of mixed Dipterocarp forest.[3] How had these people created such a marvelous *homemade* forest?

The Dayak communities in Borneo have traditionally lived in

large, communal dwellings known as longhouses. As many as thirty or forty families might share the dwelling, which is frequently elevated on stilts and, at least in the past, was made of ironwood (*Eusideroxylon zwageri*),[4] a huge forest tree with dense beautiful wood that is extremely resistant to insects, bacteria, and fungi. The front half of the longhouse consists of a large public space where the families can gather; the back is divided into bedrooms, each with its own door.

After a group of Dayak have selected a strategic spot with a good view and accessible water, they clear a small tract of forest for the longhouse. The area in front of the house is kept open so that mats with freshly harvested rice can be laid out in the sun to dry. An assortment of fruit trees, especially durian, are also planted around the house. Over time, these fruit trees will grow so tall that they start to cast shade over the open area used for drying rice. Rather than cutting down or pruning the trees, the residents dismantle the longhouse—carefully saving the ironwood boards and posts for the next house—and look for a new site to build on.

Every time a community moves its longhouse, little pockets of fruit trees remain in the forest, and over the years the members of the household will periodically revisit these trees, usually during the fruiting season, to weed, collect fruits, and plant new trees, as well as to squat on their heels, smoke, and share stories about life in the old longhouse. Over time, the clumps of fruit trees coalesce and form a small tract of forest in which almost every plant species has been planted or protected, each of the species having a particular value to the community. These small tracts of intensively managed forest orchards are known as *tembawang*. The term refers specifically to former house sites, and the majority of the indigenous Dayak groups in Borneo have been creating these species-rich, forest orchards all over the island for centuries.

When the durian trees in Bagak-Sahwa start to fruit, the entire village moves out to the tembawang. The women bring pots and pans and sacks of rice, while the children generally run around, squealing with glee at the prospect of living in the forest for a while. Meanwhile, the men get busy fixing up the palm-thatched huts from the previous year and slash weeding under the trees to make the fruits easier to find. This selective weeding, especially around the durian and illipe nut trees, is an important management tool used to maintain tembawang. During these weedings, the seedlings and saplings of valuable species will be spared, while unwanted species will be cut. And the decision—cut this one, save that one—occurs spontaneously during the swing of a bush knife.

The community stays out in the tembawang for the entire durian season, eating and sleeping in the huts, weeding and planting new seedlings, washing plates, cooking rice, and periodically, and always with an upward glance, going out to collect the durian fruits that crash to the ground.

The children are usually sent out to collect the fruits, which fall with greater frequency at night, when everyone is in the hut talking, smoking, and eating durian. At the sound of a large thud, the father will glance over at one of his children, who will grab the kerosene lamp and run out to find the fruit. The procedure sounds straightforward, but an entire village protocol lies behind it. Only certain people can harvest the fruit from certain durian trees. Families can collect only fruit that was produced by a tree planted by a relative. Old trees might have dozens of "heirs," and every one of them retains a partial right to the fruit, even those who no longer live in Bagak Sahwa. But the community looks after its own: villagers who do not have "official" collection rights to any trees will be allowed to collect a certain number of fruits from families that control a lot of trees. So when a fruit hits the ground, not everyone runs out to get it. The villagers can tell whose fruit it is by the location of the sound, and only the designated child will scurry out to pick it up. Everyone else will continue smoking and eating durian.

Many villagers eat durian each evening and toss out the husks and seeds. By the end of the season, dozens of large piles of durian husks and seeds can be seen scattered throughout the tembawang. A year later, these piles of seeds will have become dense populations of seedlings. Villagers thin out these seedlings during the slash weeding that precedes the fruit harvest, and the tallest and best seedlings will be spared to produce the next crop of durian trees. This might well be the most casual method of regenerating a tree crop ever devised.

I had several friends who were conducting anthropological research in different Dayak villages in West Kalimantan, and I was able to do botanical inventories in the tembawang at each village. The inventory sites were located at varying distances from Pontianak, in villages of differing size, ethnic makeup,[5] and degree of market involvement.

Although I had not planned it that way, the study sites offered a useful sequence for examining how differing degrees of market involvement affect the diversity and structure of an indigenous system of forest management. Current economic theories suggest that the diversity of plant species in these managed systems would decrease as the communities sold more of their products—there would be a strong motivation to plant and manage only the most valuable market species. At the highest level of market involvement, the communities would be driven to produce intensively managed plantations of a few resources, concentrating their management efforts on one or two species that generate the largest revenue stream.

The inventories in all the tembawang were made in close collaboration with local Dayak. I was able to ask them the local name of every tree, enlist their help in collecting herbarium specimens of the plant species I could not identify, and solicit information about the use of every plant resource and their reasons for maintaining it in their tembawang.[6] I was also able to leave a list of the tree species that did not bear flowers or fruits during the inventory with the villagers, so they could collect these specimens for me the next time they saw the tree in bloom.

The villages closest to Pontianak sold the greatest number of durians and fresh fruits. Farther away from Pontianak, however, illipe nuts became a more important component of local tembawang because the seeds could be dried and would remain usable long enough for villagers to carry them to the nearest road, either on foot or by motorcycle, where they could be loaded onto a truck and transported to Pontianak. Given this pattern, I was surprised to learn that one of my most remote study villages in Kapuas Hulu was growing and selling a great deal of durian fruit. They were certainly not sending the fruit on thirteen-hour trips in a truck to Pontianak. I later discovered that they were sneaking them across the border through the forest into Sarawak, where the durians were fetching an inflated price. The business skills of these Iban villagers were impressive. They would carry baskets of durian fruit across the border, negotiate a price, and sell them. But rather than simply returning with the cash and empty baskets, they would use the profits to buy cases of 7-Up, which they would sell in Kalimantan. The profit margin on the soft drink was apparently much higher than that on the fruit.

The diversity of tree species in the tembawang I inventoried ranged from 56 species per hectare to 125 species per hectare. The lowest diversity was found in the tembawang at Bagak Sahwa, which is located off a main road a couple of hours north of Pontianak. During the three years that I worked in this village, I repeatedly saw buyers from Pontianak arriving in large trucks and buying all the fruit harvested from local tembawang—durians, manogsteens, rambutans, mangos, langsats, jackfruits, everything. There was a large, thriving market for native fruits, and the farmers at Bagak Sahwa were very much involved with it. Fruiting season brought windfall profits that allowed villagers to renovate their houses, take care of sick children, buy a karaoke machine with huge speakers, send a child to college, or stuff a wad of rupiahs into their mattress. Yet they still maintained over fifty species of trees per hectare in their tembawang. Why? Their reasons varied: their grandparents had planted some of the trees, their children liked the wild, somewhat sour rambutans, the women cooked the rock-hard, green mangos. Perhaps they simply liked to collect a lot of different species in their orchards, regardless of whether or not they were worth anything in the market.

Kenyah Dayaks in the Sanggau Regency create and maintain the highest-diversity tembawang. This remote community had little involvement with the market, and the only forest product that the villagers sold was illipe nut. All the other species were used for subsistence purposes. Whatever their reason for doing so, the Kenyah are simultaneously managing 125 tree species per hectare in a forest orchard— an unprecedented feat of silvicultural prowess. In my opinion, based on what I have seen, studied, and measured over the past thirty years, these villagers are the most gifted foresters in the world.

An extension agent from the Indonesian Department of Agriculture would drop by the village of Tae a couple of times each year to talk to the farmers about improved seed, fertilizers, and the newest herbicides. He was an enthusiastic young fellow who had studied agronomy at the Institut Pertanian Bogor (IPB), reported to be the best agricultural school in the region, and he was on a mission to move local agriculture away from *primitive* slash-and-burn to more state-of-the-art, Green Revolution rice cultivation. The forests surrounding Tae contained some beautiful and well-maintained tembawang with almost a hundred species of trees, palms, and climbers. The agronomist could not avoid walking through these on his way to the rice fields. I once asked him what he thought of the sophisticated management practices employed by the Tara'n Dayaks in Tae in making these forests. He didn't know what I was talking about. The tembawang were invisible to him—just as they had been for me when I first arrived at West Kalimantan.

SEVEN

Sawmills and Sustainability in Papua New Guinea

Town of Kikori (S7°25′07″, E144°14′52″) and surrounding forests and villages in the Kikori River Basin of Gulf Province, Papua New Guinea, 1999–2002

Papua New Guinea (PNG) is one of the few countries in the world with extensive areas of tropical forests owned by local communities.[1] Among indigenous communities in the Highlands, timber extraction—a disturbing percentage of it illegal—is a key industry, although few efforts are made to develop long-term management plans or regenerate important local timber species. Villages located in the tidal swamps along the southern coast of the island, in contrast, are consistently ignored by timber companies and excluded from government forestry ventures. This is not because the

(*Opposite*) Papuan field crews heading out to run an inventory transect in the tidally flooded forests of the Kikori River Basin, Papua New Guinea. The tide was coming in; before the transect was finished, water levels in the forest were waist deep.

swamps lack forests or merchantable timber volume. Local forests exhibit a relatively simple floristic composition, contain a high density of merchantable timber species, and have excellent fluvial access to the Gulf of Papua. Rather, none of the larger timber concerns in the region wants to bring heavy equipment into swamp forests, where skidders, tractors, and front end loaders inevitably bog down in the mud and end up covered with water. As a result, no one wants to buy timber from these communities, and local forests are frequently viewed as an economic impediment instead of a resource.

In response to this situation, in 1999 I started collaborating with a community forestry project in the tidally flooded swamp forests along the southern coast of Papua New Guinea. Our goal was to provide indigenous communities in the Kikori River Basin with the technical skills needed to manage their forest resources on a sustainable basis. A small sawmill would also be set up to turn their logs into boards. We would initially train the communities to collect the inventory and growth data needed for management. If they were able to organize themselves to collect these data, we would help them develop a management plan, go into the forest with them to mark the first group of harvest trees, and pay them a premium price for the logs. Such was our plan.

The first difficulty we ran into was acquiring the sawmill. Our community forestry initiative was an extension of a previous project in which several portable sawmills had been imported into PNG and communities in the Kikori River Basin trained in how to use them.[2] When the project was over, the sawmills, for reasons unknown, had remained in the communities, which had become accustomed to cutting and selling timber. Apparently, it was never made clear that the sawmills were on loan for the project, and that they would have to be either purchased or returned at the completion of fieldwork. Considerable time and energy was invested in trying to convince the communities to return the sawmills. We finally got one of the mills back (actually one entire sawmill and pieces of a second that could be used for parts), and after several months of wrangling set it up for stationary operation on a site near the river in the town of Kikori. The mill, and the community forestry operation supplying it, was called Kikori-Pacific.

A second problem was finding villages to collaborate with us on the project. Convincing arguments can be made for sustainable resource use if forests are generating no revenue for the community, if a certain subset of the community is interested in learning new skills, and if people have the time to invest in counting and measuring trees. Community forestry is a harder sell when an oil company is paying a monthly easement fee to local communities for allowing a pipeline to cross their land. The communities getting paid this easement proved to be uninterested in going into the forest to do a strenuous job in anticipation of a *potential* source of revenue. Even the communities that were not getting paid by the oil company had hopes that a new pipeline would be built, or that another oil company would move into the area, and that they would, at some point, also start receiving monthly payments. We found it difficult to argue with this logic. After much discussion and community outreach, five village groups—Bisi, Kopi, Verabari, Komaio, and Woubo—controlling a forest area of almost twenty thousand hectares, agreed to give sustainable forestry a try.

The advantages of small-scale community forestry in tidal forests became immediately clear the first time we witnessed a crew of villagers fell a tree. To set the stage, let me first describe how mechanized, industrial logging is done in these forests. The tide has gone out, but the soil is still wet and mushy; pools of standing water dot the low-lying areas. The skidders are slipping around and making deep ruts; they periodically back into or sideswipe a tree and tear off a large strip of bark. Dragging the felled trees to a temporary landing where the logs are being stacked gouges out additional ruts and damages more trees. The whole understory has been churned into a muddy mess of roots, leaves, and shattered crowns of fallen trees—all of which will disappear underwater in a few hours when the tide comes in.

When the villagers set out to fell a tree, they walk into the forest barefoot, carrying their axes. They may have already selected and marked a harvest tree, but if not they will scout around until they find one they like. Felling the tree with an ax takes time, but eventually the tree comes down, parting the leaves of the understory sago palms (*Metroxylon sagu*);[3] the leaves immediately spring back into place to cover the gap. The villagers go home and wait for the tide to come in. They reenter the forest, this time by canoe, and tie a rope or a rattan cane around the harvest log, which is now floating in a meter and a half of water. They paddle carefully out of the forest towing the floating log behind them.

As long as no one is trying to buy logs from their forest, villagers are pretty casual about the boundaries of their property. But their attitude changes as soon as the forest acquires a commercial value. Shortly after we signed management agreements with the six communities, a number of border disputes arose and an uncomfortable amount of suspicion and infighting developed among neighboring communities. If left unresolved, disagreements among kinship groups can easily escalate into clan warfare. In hindsight, we recognized that arriving at a clear consensus about the boundaries of different communities would have been the obvious first step to sustainable resource use, especially in countries where the central government does not own the forest. At the time, however, we were caught by surprise when the community forestry work in PNG ground to a halt until the ownership boundaries of eighteen clans could be surveyed, agreed on, and plotted to scale on a 1:65,000 base map of the management area. It took several months, but we did produce such a map.

Once the boundary disputes were resolved, I gave a series of training workshops on the objectives and methodologies of forest inventory. Foresters have developed a number of ways to count trees, and some are complicated and difficult to understand. (The concept of *random* sampling, for example, can be especially tedious to explain to farmers. "Not here, not there, but everywhere with equal probability.") I have tried a variety of inventory methods over the years, with mixed results, and have come to the conclusion that the simplest is decidedly the best.

Years earlier, in Borneo, I was involved in a large forest assessment project designed by a German inventory specialist. We located random plots on satellite photos of the area and had GPS devices to help us find them.[4] The sampling scheme was high-tech and statistically rigorous. We would head out from the village in the morning, walk several hours through the forest, arrive at the plot location, and spend the next twenty minutes counting and measuring a few trees. We would then pack up the equipment and start hiking to the next plot, which usually took several more hours. By the end of the day, we had done only two or three inventory plots, but had walked for about six hours. One of the Dayak field assistants complained that he could not understand why we did not just count trees while we were walking. Good point. I have used long transects (a series of contiguous, narrow plots oriented systematically throughout the sample area) for all my community forestry inventories ever since.

I had no difficulty explaining how to run a transect to the village field crews. With a compass, one person first selects an appropriate heading, and then stretches out a twenty-meter length of rope in that direction. The person pulling the rope clears a little path with his bush knife; the rope is left on the ground to provide the centerline of the plot. Two villagers, one on the right side and one on the left side of the line, then start searching within a five-meter-wide strip on either side of the centerline for the tree species on the inventory list.[5] Once a tree is found, its diameter and height are measured and notes recorded about the overall form and health of the tree. These annotations are especially important. Finding a timber tree with a diameter of forty centimeters and a height of twenty-two meters is good news—unless the finder's notes include "trunk is crooked and has a large burl on one side." The crew continues searching on either side of the line until the men reach the end of the rope. The compass is used to check the bearing again, and the rope is stretched out another twenty meters for the next plot. And so it goes, plot after plot, until the crew arrives at the boundary of the property.

The inventories conducted by the six villages focused on three major timber species—mangrove cedar (*Xylocarpus granatum*), *kwila* (*Intsia bijuga*), and *kalofilum* (*Calophyllum papuanum*)[6]—and working with Kikori-Pacific foresters, the villagers counted and measured all the trees of these three species in almost sixty kilometers of inventory transects. Relative to the quality and quantity of the inventory work conducted by the major timber companies in PNG, this was a major accomplishment. There were, however, a few difficulties to overcome in persuading the communities to do the fieldwork. Several of the villages initially asserted that they would not go to the field unless provided with transportation, per diems, and fuel, and a few others required the continual assistance of Kikori-Pacific foresters to complete the transects.

One day a group of landowners from the village of Gibi showed up at the mill and said they wanted to join the project. We checked on a map and saw that their village was located far downriver in the Kikori estuary on a little finger of land sticking out in the Gulf of Papua. When we asked whether they had a boat to pull logs, they shook their heads. We had doubts about their ability to transport sawlogs to the mill, and were reluctant to commit to the training and inventory work if neither the mill nor the community was likely to benefit. As a compromise, we suggested that they return to their village and fell some trees. If they were able to get the logs to Kikori, we in the project would be happy to collaborate with them. After they left, we promptly forgot about the incident.

Two weeks later, the villagers showed up again. They had gone back to Gibi, felled a dozen trees, and tied them together to make a raft. They floated it up the Kikori River each day with the rising tide as far as they could, and when the tide started to recede they tied up next to the bank and waited for the river to start rising again the next morning. It took them five days to get to Kikori, sleeping on top of their logs. We gave them something to eat and enthusiastically welcomed them to the project.

After three years, Kikori-Pacific was employing twenty-four local people and had a forestry staff of five technicians. There were over two hundred landowners from nine villages collaborating with the project. These landowners had inventoried their forests, carefully selected and marked harvest trees according to a management plan, felled the trees with axes, and floated the logs to the mill at Kikori. The mill had purchased thousands of sawlogs, and paid hundreds of thousands of kina (the local currency of PNG) to participating villages. But the mill was hemorrhaging money. The project probably grew too fast, involving too many communities and acquiring too many logs, which piled up at the mill. In addition, finding markets for the wood posed new problems. We could sell rough sawn boards to the local oil company for temporary housing and storage buildings, or to nearby communities for schools and public buildings, but only at a low price. To obtain a better price for the wood, which was sustainably harvested and of unquestionable quality, we would have to export it. Shortly after the initial funding for the project was depleted, the mill was forced to close.

We can take many lessons from this project. The first, fundamental one is that it is extremely difficult to work in the tidal swamps of Papua New Guinea. The local political and social situations are complex, the long-term presence of a multinational oil company creates unfilled expectations for communities, and the daily rise and fall of the floodwaters can make fieldwork exhausting and tedious.[7] Had we chosen to export the wood, the paperwork and permits required would have taken some time to arrange, and we did not consider it worthwhile to go through the complicated process because we did not have a kiln to dry the boards or the money to buy one. And no one was optimistic about the prospects of exporting *green* boards from PNG to another country.

Timber operations in PNG are thus stymied by the general failure to involve communities in such a way that the timber is harvested sustainably, the wood resources are conscientiously managed, and the value-added opportunities are economically viable. Until procedures for these are set in place, local communities will continue felling trees indiscriminately for whoever wishes to buy the logs, without regard for sustainability, resource management, or even profitability. The Kikori-Pacific experiment took small first steps toward a different goal. We proved that villagers who live and work in the forests of PNG can conduct forest inventories and develop sustainable management plans as well as university-trained foresters—if given the opportunity. The future of sustainable forestry in PNG will depend on whether local communities continue to be given these opportunities.

EIGHT

Collaborative Conservation in the Bwindi Impenetrable Forest Reserve

Bwindi Impenetrable Forest Reserve, Kabale District, Uganda, 2001

In 2001, I was invited to the Bwindi Impenetrable Forest Reserve in southwestern Uganda to co-teach a short workshop for the wardens and technical staff of local protected areas. The purpose of the workshop was to provide some of the analytical tools necessary to facilitate a greater collaboration with local communities in the conservation and management of the parks. Our plan was to talk a little about ecology, ethnobotany, and resource management, go into the field to survey some of the plants that were important to local communities, run a few inventory transects, and assess the impacts—or lack of impacts—that the current rate of harvest was having on different resources.

(*Opposite*) Female gorilla (*Gorilla beringei*) in the Bwindi Impenetrable Forest Reserve, Uganda.

The setting and complicated community resource-use situation made this workshop different from many of the others that I had given. Bwindi is not a large tract of forest with a low-density human population such as the ones I visited in Borneo and Amazonia. Rather, it is situated in one of the most densely populated areas of Uganda and is an island, rather than a sea, of forest surrounded by farmland. Agricultural fields stretch right up to the boundary of the park, and the distribution and abundance of many of the valuable forest species show the effects of decades of pit sawing and the uncontrolled harvest of trees for beer boats (carved wooden troughs used for brewing banana beer), firewood, building posts, and bean stakes.[1] About half of all the mountain gorillas (*Gorilla beringei*)[2] in the world live in the Bwindi Impenetrable Forest.

Harvesting by villagers in the Bwindi forest is tightly controlled and monitored by the park wardens, and it is a frequent point of contention between the rural communities and the conservation groups that fund and manage the park. Gorillas' needs, the needs of the local farming communities, and the regeneration and growth requirements of the native flora all have to be respected to maintain the forest and achieve the long-term conservation objectives of the reserve. Through the discussions and field exercises, we hoped to find common ground during the workshop for the people, the plants, and the gorillas.

As we gathered outside for dinner after a lively discussion about the origin of economic plants, a couple of students started making comments about the food on their plates. The assertion that tomatoes came from Uganda was met by the rejoinder—from me—that tomatoes were originally from Central America. Another student chimed in that chickens came from Uganda; well, no, chickens come from jungle fowl native to Southeast Asia. But these delicious mangos are from Uganda, aren't they? Nope, they are also native to the forests of Southeast Asia. At this point, the students were becoming a little frustrated and testy. Finally, one of the students asked, "Well, what *does* come from here?" My ethnobotanist colleague Tony Cunningham, who had organized the workshop and was an old hand in East Africa, thought for a moment, and then in a quiet voice said, "Human beings."

Humans beings have occupied the Bwindi region for over 37,000 years. The first evidence of forest clearing dates back 4,800 years, when small populations of hunter-gatherers used fire to manipulate the vegetation. These groups hunted and harvested a variety of plant resources from the forest, casually planting the seeds of a few species in the ashes following the burn. The first wave of dedicated farmers arrived at the region about 2,000 years ago. The indigenous communities currently farming near the boundary of the park—and occasionally wild harvesting plants inside the park—have extensive knowledge about the plants and animals that occur in the Bwindi Impenetrable Forest. Their cultural traditions reflect deep relationships with the forest and critical subsistence dependencies on local plant species; they have coexisted with gorillas for thousands of years. These relationships and dependencies do not disappear when local communities are denied access to the forest. As we repeatedly emphasized in our workshop, much could be learned by including local communities in a discussion about how best to manage and conserve the gorillas and the plants at Bwindi.

A small population of African elephants (*Loxodonta africana*) lives in the forest at Bwindi.[3] A couple of times while running our inventory transects we passed through areas where a family of elephants had been feeding. African elephants eat several hundred kilograms of vegetation each day, and their feeding areas look like bombsites, as this one did. All the leaves had been stripped off the understory trees, pole-sized stems had been broken in half, and several of the larger trees had gashes where the bark had been ripped off and the trunk scraped by tusks. A couple of trees had been completely pushed over and uprooted, perhaps by one of the elephants accidentally backing into it. It must have been quite a meal.

Communities of Bachiga, an ethnic group from northern Rwanda and southern Uganda, settled near the park weave the stems of two herbaceous species, *Eleusina indica* and *Plantago palmata*,[4] to make baskets. Both these plants are common in disturbed areas and along the side of the road, and in many countries they are considered weedy, invasive species that need to be controlled. But perceptions are different in a conservation area. At the time we conducted our workshop, the harvest of *any* type of plant material within the park was forbidden, and local weavers were prohibited from picking a handful of wiregrass to weave a millet basket.

One species of liana (*Loesneriella apocynoides*),[5] known as *omujega*, is an important source of fiber for local weavers. The stems are extremely durable and resistant to rot and attack by wood-boring insects, and they can be dried and stored for years until they are needed. Because of these properties, omujega is harvested and used in large quantities to make granaries and ambulance stretchers, and for general basketry. The recent development of the tea industry around Bwindi has further increased the demand for omujega fiber to make baskets for harvesting tea. This climber, however, grows very slowly and occurs at low densities in the forest. We encountered only a couple of individuals in our transects, and all were sprouts from larger stems that had been harvested. The current situation of this species is tenuous. It is growing in a protected area, but that in itself does little to guarantee the plant's continued existence.

Every type of resource use in the park, either by elephants or Bachiga weavers, creates a disturbance that has an impact on the composition and structure of the forest. The current forest at Bwindi was produced by hundreds of years of sporadic plant collection by local villagers and the daily foraging behavior of elephants.

Prohibiting one type of use that might have a relatively minor ecological impact to meet certain conservation objectives while allowing a greater source of disturbance to continue for the same reason reflects a general misunderstanding of forest ecology. To conserve a forest in its current state and maintain a stable habitat for the gorillas, the original disturbance regimes that have created that forest must also be conserved. Restricting the traditional harvesting practices of local communities will change, not conserve, these patterns. Building a more obligate relationship between the resource users and the resource would seem to offer a better solution. Train the weavers and plant collectors and put them in charge of doing the resource inventories. Based on what they find, let *them* set—and enforce—the harvest quotas. These are the types of collaborative conservation alternatives that we were trying to address in our workshop.

After we finished the workshop, the park managers graciously allowed us to accompany a team of Batwa trackers who had been following a group of gorillas for several months. (The Batwa people are one of the last groups of pygmies in Uganda. They lived as hunter-gatherers in the Bwindi forest until they were forced out in 1964.) The trackers were working to gradually *habituate* the gorillas so that they could be observed and studied more closely by scientists. The trackers would look for signs of the band and wander around through the forest until they found the gorillas. After they found them, the trackers would spend a little time with the animals, stopping a considerable distance away but close enough for the gorillas to be aware of their presence. The trackers would return the next day and get a little closer. Again, they would do nothing but squat down in the forest for an hour or so and then quietly walk away. After the trackers had visited for a few weeks, the gorilla band would start to ignore them, even if the trackers squatted only two or three meters away from female gorillas with babies or a large silverback male. I appreciate the potential scientific benefits of habituation, but I question the logic of gradually training gorillas not to be afraid of the one thing that they should definitely fear—human beings.

After several hours of hiking with the Batwa trackers, we located the study band and were able to get uncomfortably close to them. The Batwa showed us how to squat down and make little grunting noises while we pretended to eat grass. We were told never to look up and make direct eye contact. I did *exactly* as I was told. On one hand, I could not believe I was so close to some of the most elusive, threatened, and noble creatures on earth. On the other, I was terrified. The gorillas moved around as they ate, and I tried my best to keep a tree between the silverback and myself.

How it happened I don't know, but after a few minutes I noticed that I was positioned precariously with all of the females and the babies on one side of me and the silverback on the other. I was still squatting next to a tree, but in a frighteningly inconvenient place. I continued to grunt and pretend to eat grass. And then the silverback lunged in my direction, grazing the tree I was squatting next to with his shoulder, and took up a position next to one of the females, where he casually started to eat again. He probably had not even noticed me. I sure noticed him.

The development of community-based resource management activities at Bwindi would achieve several conservation objectives. Allowing villagers to continue traditional patterns of resource harvest in the forest, but on a sustained-yield basis as determined by inventory and growth data that they collected themselves, would promote forest conservation and reestablish a sense of forest stewardship in local communities. Having more people in the forest who support the conservation agenda of the park would also act as a deterrent to poaching and illegal harvesting. The greatest benefit, however, would be the maintenance, and ideally improvement, of quality forest habitat for mountain gorillas. The first step, at Bwindi as well as in many other protected areas in tropical forests, is greater involvement of the people who live along the perimeter of the park.

Eco alebrijes

NINE

A Renewable Supply of Carving Wood

San Juan Bautista Jayacatlán (N17°25′37″, W96°49′11″), Central Valley of Oaxaca, Mexico, 2001–2003

Alebrijes are small, whimsical figures that are carved and painted by Mexican artisans in the form of animals, dragons, mermaids, or human-animal hybrids. They are not a traditional craft item but rather a toy that farmers in the Central Valley of Oaxaca would carve for their children. Craft buyers started to notice—and buy—alebrijes in the early 1960s, and soon the carvings were appearing in craft stores throughout Mexico. By the 1980s wholesalers from the United States were visiting the carving communities and purchasing the figures directly from the makers. The alebrijes became more intricate and the painting more ornate as the carvers sought to distinguish themselves in an increasingly competitive market.

(*Opposite*) Alebrije produced from *copal* (*Bursera glabrifolia*) wood sustainably harvested from the dry forests of San Juan Bautista Jayacatlán, Oaxaca, Mexico.

By the 1990s, most of the households in the better-known carving communities such as San Antonio Arrazola, San Martín Tilcajete, and La Unión Tejalapam were earning a significant portion of their annual income from the sale of alebrijes, and some families had abandoned agriculture to work as full-time carvers. As the market continued to grow, it became apparent that individual families could not keep up with the orders using only household labor, and several alebrije factories, each employing twenty or so workers to help with the sanding and painting, were established in the vicinity of Oaxaca City.

But as the demand for these figures boomed, the supply of the raw material required to make them diminished. All alebrijes are made from the wood of a single tree species (*Bursera glabrifolia*),[1] native to the tropical dry forests of Mexico. Known locally as *copal*, this tree produces a soft, fragrant wood that is ideal for carving. Unfortunately, given the relatively harsh conditions of a tropical dry forest, copal trees do not grow very quickly and their regeneration is limited. Tropical dry forest habitats exhibit high temperatures, low total annual rainfall, and a marked dry season of from five to eight months. These are not ideal conditions for plant growth. Population densities of the species are thus low in many regions. As well, the species has been exploited since pre-Columbian times for its aromatic resin and the fragrant oil found in its wood. Copal trees in Oaxaca are struggling.[2]

Once the market for alebrijes began to grow, the demand for copal wood skyrocketed and all the copal trees in the vicinity of the carving communities in Oaxaca disappeared. Outside wood collectors entered the picture and started bringing illegally harvested copal to the carving communities—usually at night—and selling the material for exorbitant prices. Every year the wood collectors were forced to go farther into the mountains to look for copal trees, and the selling price for the wood correspondingly increased. The carvers, however, continued to receive the same price for their alebrijes.

The situation highlighted two related problems. Because commercial tree felling in Mexico is illegal without a permit and an approved management plan—especially in government-controlled protected areas—copal wood was being harvested illegally from dry forests, mostly in nearby protected areas. The species was being overexploited and totally depleted from increasingly larger areas of the Central Valley of Oaxaca. Yet in spite of a burgeoning international market for their products, local carvers were trapped in a vicious price squeeze and earning less and less for their work. This familiar pattern illustrates how commercial markets for wild-harvested resources move from the *boom* to the *bust* phase.

In an effort to reverse this downward spiral of resource depletion, I collaborated with a Mexican colleague, Silvia Purata, on a project focused on the sustainable use and management of copal trees in the tropical dry forests of Oaxaca. We reasoned that such a project would produce several useful results. It would provide an immediate source of carving wood, enhance the value of dry forests relative to competing land uses such as agriculture and pastures, and, we hoped, offer an incentive to conserve these habitats. Perhaps most important, however, the project would engage local communities in the long-term management of their forests and offer a mechanism to compensate them for their efforts.

Our first task was to find the right community. We needed a little town that was well organized, was motivated to try a new approach, and had control of several thousand hectares of tropical dry forest. After visiting a couple of communities in the mountains outside of Oaxaca, my colleague and I finally fixed on San Juan Bautista Jayacatlán, which seemed perfect. During our first meeting to explain to the community what we had in mind, we became a bit overexcited about the prospects for sustainable forest use and rapidly outlined how we would lay out a management area, help the villagers do an inventory of copal trees, start growth studies, write a management plan and submit it to SEMARNAT (Secretariat of the Environment and Natural Resources, the ministry that has to approve a management plan and grant a permit before a community can harvest and sell trees) to get a harvest permit, and then sell the sustainably harvested wood to the carving communities. At this point, the village head, who had been very patient with us, commented "No me acelera, maestra"—Not so fast, Professor, the community has to approve all of these things first.

At that point we left so the villagers could confer among themselves; they had another meeting or two, and before much longer, we were informed that the community wanted to start managing the copal trees in their dry forests and selling the wood to the carving communities.

I had a gifted and energetic research assistant at this time, Berry Brosi, who agreed to move to Oaxaca for the summer to set up the management area and help the community with the copal inventory. Mariana Hernández-Apolinar, a doctoral student at the National University of Mexico, had also decided to collaborate with the project and conduct a population study of the copal trees to assess the ecological sustainability of harvesting. And a local forestry group had been working with the community to help the villagers manage their pine timber.[3] Of course the villagers themselves were prepared to assist with the fieldwork. During the summer of 2003, this combined team collected an enormous amount of baseline data about the copal trees within the management area. We learned how many copal trees there were in the forest, how big they were, and where they were located. We laid out seedling plots to determine how well the species was regenerating on the site, and put growth bands on trees of different sizes to see how fast they were growing.

Estimating the wood volume of the copal trees took some extra calculations. The volume of most timber trees can be estimated relatively simply by means of the diameter and commercial height (the height to the first major branch or fork along the upper stem) of the tree. Copal trees, however, are short and have a lot of branches, and these branches are the most desirable type of wood for the artisans who carve alebrijes. We came up with a rather ingenious way to calculate the wood volume of a copal tree by felling a few trees of different diameters, cutting the stems and the branches into sixty-centimeter-long pieces, and then weighing each one with a spring balance. We next summed the weights of all the individual pieces to calculate the total weight in kilograms of each tree. Finally, we converted the weight measurements to volume using the estimated density or specific gravity (grams per cubic centimeter of wood) of copal.[4] We made a graph showing the total wood volume for trees of different diameters and used this to calculate the volume of every copal tree in the management area. It was simple, it was elegant, and it told us how much copal wood the community had in their management area. The growth bands recorded how much new wood was being produced each year.

Using these data, together with other information that she had collected, Mariana Hernández-Apolinar made a sophisticated computer model to assess the impact of wood harvesting on the regeneration and growth of the copal population, and ran several simulations to determine how much wood could be harvested on a sustainable basis. She found that the copal population was increasing slightly each year, and that the population could sustain an extraction rate of about eight trees per hectare per year. In a relatively short period of time, we had compiled an exhaustive dataset about the structure and growth of copal trees in this tract of dry forest.

It was now time to put all these data together into a management plan and submit it to SEMARNAT to get a harvesting permit for the community. Berry Brosi did most of the work on the plan, and the result was thorough, extremely well documented, and rich with charts and maps and regression equations. I remember thinking that it had to be one of the most detailed management plans ever written for a tropical dry forest in Mexico.

We submitted the proposal and then waited. And waited. We finally gave up and called SEMARNAT. We first asked the official if he had received our management plan and application for a harvest license; he had. We then asked whether there was a problem with the management plan, or if we had made a mistake on the application; no, everything looked good. (In fact, he complimented us on the quality of our management plan.) After several minutes of small talk, the SEMARNAT official confessed that he had no idea what to do with our application. Because of the small size and poor form of the trees, tropical dry forests contained little timber of value. No one had ever submitted a management plan for harvesting wood in a dry forest before. Ever. He then confided that people do not manage dry forests in Mexico; they generally convert them to agriculture or cattle pastures. But we kept him on the line while we made the case for sustainable forest management as a way of conserving tropical dry forests in Mexico. By the time we hung up the phone, our proposal had been approved.

The harvest of the first copal tree under the permit was conducted with a fair amount of ceremony. Everyone walked out to the field with the chainsaw, the special hammer to mark the stump of the tree,[5] a bottle of cheap whiskey, and some plastic cups. We selected the first copal tree, hammered the stump with the seal, filled a cup with whiskey, and offered the following toast: "With the permission of the forest, and the owners who are here with us, we offer a toast to start work on the management and exploitation of the copal resources found here." We then poured the whiskey around the base of the tree, and the sawyer yanked the starter cord on his chainsaw. It emitted a little puff of smoke, and down came the first sustainably harvested copal tree in Mexico.

After years of buying illicit copal wood at inflated process, carving communities were thrilled to hear our sales pitch for sustainable harvesting and placed their orders. The wood would not be delivered in the middle of the night, and it would be significantly cheaper than that offered by other sellers because it came from a forest that was closer to Oaxaca—and it was legally harvested. We also tried to give the carvers precisely the type of wood they wanted. If they wanted thin, twisty branches, we would give them thin, twisty branches. If they wanted large trunk pieces free of knots, we would try to deliver those. We could not produce enough wood to supply all the carving communities in the Central Valley of Oaxaca, but we set a new standard of service for the ones we could reach, and provided a model for sustainable copal harvesting in the tropical dry forests of Mexico.

Expresso
Cargas

TEN

Caboclo Forestry in the Tapajós-Arapiuns Extractive Reserve

Tapajós-Arapiuns Extractive Reserve, western Pará, Brazil, 2002–2005

The logging companies, local business interests, and municipal authorities never wanted it to happen, but in 1998 a 650,000-hectare block of forest between the Tapajós and the Arapiuns Rivers in western Pará was turned into an extractive reserve, a controlled area in which the state allocates use rights, including resource extraction, to local communities. The *caboclo* communities—detribalized groups of mixed European, African, and indigenous descent that inhabit the floodplains of lowland Amazonia—living along the river had been fighting for years to oppose commercial logging in the forest, and aided by mounting political pressure, they had finally won. Three kilometers of land paralleling the rivers along the perimeter of the

(*Opposite*) Woodworker from the village of Nuquini in the Tapajós-Arapiuns Extractive Reserve, Pará, Brazil, sculpting the back of a handcrafted chair. Photo by D. McGrath.

area was designated for settlements and agricultural activities; the remaining land was gazetted as a forest reserve. Resource use in the Tapajós-Arapiuns forest, one of the first examples of a *new* type of extractive reserve being created in Brazil, was not limited to rubber tapping and the collection of Brazil nuts. Communities were not allowed to sell whole logs, flitches (cross-sections), or rough-sawn timber from the reserve, but when value-added processing—and a written management plan—could be provided, they were permitted to fell a specified number of trees each year.

Three of the caboclo communities in the reserve, Nova Vista, Nuquini, and Surucuá, expressed an interest in developing a small, handcrafted furniture initiative. They had been making a few stools and tables using wood from dead trees in the agricultural zone, but to ramp up production and access timber growing in the forest reserve, they needed a formal management plan and a permit from the Brazilian Institute of the Environment and Natural Resources (IBAMA). Since they did not know how to go about doing this, I started collaborating with these communities in early 2002.

During my first meeting with the communities, I explained that we needed inventory and growth data from the reserve to develop a management plan. Without going into too many details, I described how we might collect this information. We made a list of the forty to fifty species of desirable furniture woods in the reserve and had a long discussion about sustainability and how forests could be conserved through wise use and compliance with the rules. As we were wrapping things up, I asked the representatives from each village how much of the extractive reserve they were thinking of including in their management plan. I was expecting them to suggest one or two hundred hectares of forest for each village—after all, they were making cutting boards and stools, not feeding a sawmill. The chatted among themselves for a minute, and then told me in a serious tone that they wanted to manage all of it. All 650,000 hectares. That's a lot of stools.

I spent some time explaining what was involved in a forest inventory, after which it was unanimously agreed that a two hundred hectare intensive management area would produce more than enough wood to keep their furniture initiative going. (After the data from the inventory had been collected and analyzed, we calculated that one tree with a diameter of 1.5 meters and a height of 45 meters—of which there are many in the management area—contained enough wood to produce about fifty thousand cutting boards.) We started the inventory work in Nuquini. The management area was located about four kilometers from the village, and rather than walk there twice a day—especially in the late afternoon when we were exhausted—I suggested that we camp out in the forest until we finished. Nobody liked this idea. A group of women from the village agreed to make lunch for us each day, but we always had be back in time for dinner each night.

We first laid out a baseline of stakes every hundred meters along the southern boundary of the management area to mark the starting points of the inventory transects. We then divided ourselves into two teams and started slowly moving north using a compass and measuring tape, counting and measuring all the trees we encountered on the species list that were larger than five centimeters in diameter. Each of the transects was a kilometer long, but neither team got even half through the work the first day. Constant questions arose about using the clinometer, an optical device used to measure slope or tree height, determining whether a tree was in or out of the transect, measuring the diameter of a large tree with buttresses, and writing the data in the correct column on the tally sheet. The wasps' nest in transect 1 that sent everybody running into the forest did not help things.

We grew progressively faster at the work as the days went on, and by the end of the inventory, the crews were doing a transect and a half or more a day. We were proud of what we had accomplished, and we had gained a much better idea of the distribution and abundance of furniture woods in the management area. Most evenings were spent animatedly sharing this information with the rest of the village, whether anyone wanted to hear it or not. The inventory work at Nuquini took a total of eight days to complete. Two teams of nine extremely motivated villagers sampled 1,000 plots, counted and measured 3,452 trees from 42 species, and encountered 5 poisonous snakes (*Bothrops* spp.) and one very large red-tailed boa constrictor.[1]

We had heard about the abundance of poisonous snakes in the forest and were determined to have a good supply of antivenin for the crews. This turned out to be more complicated than we had expected. We could not simply buy the antivenin and some disposable syringes and take them to the field. We had to contract with a nurse to administer the antivenin, and he would have to accompany us to the field carrying an ice chest. The young man we hired for this was wonderful, and after several days without having to open his ice chest or give anybody a shot, he started helping us count and measure trees. No one was bitten by a snake, but he was certainly prepared: he had fifty doses of antivenin in his ice chest—enough to treat up to six cases of snakebite a day for the entire inventory.

Several young women from Nuquini also helped with the inventory. They had the best handwriting and were the most careful about recording the measurements in the correct columns, so they were always chosen to carry the clipboard and tally the data. All these women had families and were expected to help with the dinner preparations at their houses each evening. I will never forget the sight of one of our most capable and trusted note takers handing her clipboard to another member of her team one afternoon so that she could pick up a large tortoise (*Chelonoidis denticulate*)[2] that was slowly walking through the transect. She flipped the tortoise over on its shell, and carried it around, much like a waiter carrying a tray, for the rest of the day. Dinner solved.

I wanted to train the villagers to make dendrometer (growth) bands, which they could leave in the forest and periodically check to measure the diameter growth of important furniture woods. These bands are relatively easy to make, and since growth data ultimately determine how much wood can be harvested each year, I considered it important for the villagers to collect these data themselves. The government foresters I talked to, however, told me that extensive growth studies had been done several years earlier on the other side of the river in the Tapajós National Forest by *university-trained foresters.* They recommended that I use these data. There was the unstated, but nonetheless palpable, opinion embedded in this comment that the villagers would not be able to make the bands or do a growth study by themselves.

The last thing I wanted to do was simply give the community a number derived from somewhere else—like pulling a rabbit out of a hat—of the annual growth of their trees to combine with the inventory data they had worked so hard to collect so they could calculate an annual allowable harvest. They deserved the opportunity to take responsibility for gathering these data for themselves. Another advantage of having the villagers put growth bands on the trees in the management area was that the community could then measure the growth response of the residual stand after the harvest trees were removed. As the canopy was opened through tree felling, the remaining trees should grow more quickly. Faster-growing trees would mean greater wood production each year and larger volumes of wood that could be sustainably harvested from the forest.

It took a single morning to train the villagers to make the dendrometer bands. And once they learned how to make the bands, they wanted to put them on as many trees as possible. We ended up banding more than 250 trees from a variety of species, and the results from the first year of growth were invaluable. All the tree species measured in the management area were growing significantly faster than the same species across the river in the National Forest. A forest fire had swept through the area several years earlier. The fire had quickly been suppressed in the National Forest, where personnel and equipment were available to deal with it; it burned for a considerably longer period in the forests behind Nuquini. As a direct result of the fire, the canopy was more open in Nuquini's management area than in the National Forest, light levels were higher, and, not surprisingly, the trees were growing faster. Everyone in the village was delighted with this finding.

Once the inventory and growth data were available, we had several productive meetings playing around with the data and assessing different scenarios for determining the sustainable annual harvest of wood from the management area. To do this, we first needed to calculate the wood volume of every tree recorded in the inventory, then group these data into species, expand the diameter of each tree using the average annual diameter increment of that species, and finally recalculate the volume based on the new diameter. By subtracting the original wood volume from the new volume calculated after a year of growth, and figuring out the sum of all these, we could estimate how much new wood a particular species had produced in one year. This result represents the inviolate limit of how much wood can be harvested from that species each year on a sustainable basis. It may not be rocket science, but forest management *can* get a bit complicated. I should point out that every member of the forestry team at Nuquini, some of whom could not read, stayed with me every step of the way as we did these calculations. There was much whispering back and forth as I wrote things on the blackboard, but based on the responses I got when I questioned them, the caboclos understood the procedure very well.

Combining the inventory and growth data yielded some fascinating results.[3] For example, one cubic meter of wood contains about 1,000 cutting boards, 190 stools, or 50 coffee tables. Or, expressed in terms of growth and sustainable harvesting, the two-hundred-hectare management area at Nuquini produces 184,000 cutting boards, 35,000 stools, and 9,200 coffee tables each year. This bodes well for the sustainability of the venture, because these quantities exceed by a huge margin what the furniture makers can produce—

or sell—each year. Suffice it to say that these figures leave a lot of room for the small furniture initiative to grow.

After finishing the inventory work at Nuquini, we moved to Surucuá and started counting trees in another piece of forest. The management area selected for the inventory was located ten kilometers from the village, but this time the field crews decided to build a basecamp so we could sleep and have our meals in the forest. The first day, a long line of us headed down the trail to the management area like a stream of ants: workers carrying pots and pans, sacks of food, hammocks and tarps, bags of field equipment, and a boombox. Everybody was on a bicycle. It was strangely reassuring to come back from the field each evening and see twenty-five bicycles parked next to the basecamp. My bike was bright red, all of the padding had come out of the seat, and the brakes didn't work.

When I was not sleeping in a basecamp, I spent most of my evenings in the Tapajós-Arapiuns Extractive Reserve on a small boat with colleagues from the research institute in Santarem that was hosting my visit. We would take our meals on the boat, eating off a metal plate while rocking in a hammock, and then climb down off the boat into the river, usually in the late evening, to bathe and reflect on the day's events. And this may be the best-kept secret of the reserve—white sand beaches and no bugs.

A final note: the furniture makers do not use large logs or long boards in their work; they prefer to work with small "bolts" of wood because these are easier to handle. Material of this size can be easily transported out of the management area across the seat and handlebars of a bicycle. The forests are essentially flat, with few rocks, and trails running all over them, so everybody has a bicycle. In a wonderful nod to combining logging technology with conservation of fossil fuels, the harvest trees at Nuquini are felled with a chainsaw and bucked into one-meter-long sections that are then wheeled out of the forest on a human-powered bicycle.

ELEVEN

Measuring Tree Growth with Maya Foresters

Eight forestry *ejidos* along the eastern coast of the Yucatán Peninsula in Quintana Roo, Mexico, based in Chetumal (N18°30′13″, W88°18′19″), 2005–2007

The Selva Maya extends through Guatemala, Belize, and the Yucatán Peninsula of Mexico. Comprising more than five million hectares, it is the largest contiguous tract of tropical forest in Central America, second only to Amazonia in the Western Hemisphere. Much of the intact forest in Amazonia is conserved in parks, extractive reserves, and other types of protected areas under government control. In contrast, over half of the forests in the Selva Maya are owned by *ejidos*, communities that collectively control, use, and manage their agricultural and forest lands.

Local communities have been managing forests in the Selva

(*Opposite*) Maya forester putting a growth band around a *tzalam* (*Lysiloma latisiliquum*) tree in the Selva Maya of Quintana Roo, Mexico.

Maya since pre-Columbian times.[1] In recent years, community forestry activities in the region have largely focused on the production of timber—in particular, export-quality mahogany (*Swietenia macrophylla*).[2] The forestry operations of several ejidos in the region have been certified by the Forest Stewardship Council (FSC) to be sustainable, and some have been able to maintain their certification for more than twenty years. These are some of the oldest certified tropical forests in the world.

Many ejidos in the Selva Maya are facing a critical period in the conservation and management of their forests. Several are coming to the end of the rotation period (the length of time between establishment and harvest; it usually includes a series of intermediate cutting cycles) and require new management plans, and others have been given a detailed list of deficiencies that must be corrected before their certification is renewed. In each case, the stumbling block is lack of basic information about tree growth. These data are not expensive or difficult for villagers to collect, but most rural farmers do not know how to make a dendrometer band or conduct a growth study. The growth estimates used in the management plans of many ejidos, for example, are based on government research conducted in other parts of the country, which may not be representative. If the growth estimates are too high, the allowable cuts calculated in the management plan will also be too high, and more wood will be harvested than the forest can sustain. If the growth estimates are too low, valuable timber that could be harvested sustainably will be left in the forest. The best management plans are those based on growth data collected from trees in the forests being managed.

I started working with a group of eight forestry ejidos in 2005 to initiate growth studies on local timber species. In addition to mahogany, the ejidos monitored the growth of twenty important tree species; almost three thousand trees were fitted with dendrometer bands. This might have been the largest community-based study of tree growth ever initiated in the tropics.

We held three workshops in Quintana Roo to train ejido foresters to conduct a growth study. The format of all three workshops consisted of a morning classroom session focused on the basic principles and methodologies of measuring tree growth, followed by an afternoon session in the field to make dendrometer bands and teach the villagers how to use the other instruments required for data collection, such as a vernier caliper to measure the growth increment recorded by the dendrometer band; a 10-factor basal area prism to count the number of neighboring individuals (competitors) in the vicinity of each sample tree; and a spherical densiometer to quantify the percentage of canopy cover. Each ejido drew up a list of the important timber species that grew in the community's forests, and drafted an agreement with the project to formalize its commitment to band and monitor the growth of a certain number of sample trees.

Participants in the three workshops responded similarly to the basic forestry concepts and tools we introduced. Foresters have for many years measured tree growth with a diameter tape (d-tape): a steel or cloth tape that expresses the measurement of the circumference of a tree in terms of diameter by dividing it by pi. The diameter of a sample tree is measured with the tape, a specified length of time is allowed to pass, and tree is measured again. The difference between the two measurements is assumed to represent growth. The inherent problem with this method is that many tropical trees grow extremely slowly, and measuring tree diameter with a d-tape is subject to many sources of error depending on where the tape was positioned around the trunk, how tightly it was pulled, whether the bark was cleaned before the measurement, and whether the tape was read correctly. All these vary depending on the person making the measurement. To demonstrate the difficulties, during the morning sessions we would select a sample tree and ask each person to measure its diameter with a d-tape and write the result on a piece of paper—but not tell anyone what it was. As expected, the results differed significantly, in some cases by as much as one or two centimeters. If the annual diameter increment of a timber species is only half a centimeter per year, the d-tape method is not going to provide useful information. Given the amount of grimacing and animated conversation, usually in the Mayan language, that invariably occurred when we compared the diameter measurements from the sample tree, all the workshop participants were convinced that a better way of measuring tree growth was needed.

A similar degree of fascination and enthusiasm was displayed when we showed them how to make dendrometer bands. Each ejido team was given a plastic lunchbox containing several rolls of stainless-steel strapping, several dozen springs, a pair of scissors, a permanent marker, and a vernier caliper, and was promised that more material would be on its way after we had had a chance to review their species lists. In each of the workshops, for every team from every ejido, once we showed them how to make a dendrometer band, it was impossible to get the lunchbox back. The dendrometer band is a simple, yet elegant tool, the utility of which everyone immediately appreciated.

Some ejidos contracted to band more trees than other communities did, but all the communities agreed to start collecting data on the growth of the timber trees in their forests. Even the smallest ejidos committed to band at least three or four hundred trees.

There is a machine shop in Connecticut that makes stainless-steel springs. Shortly after returning from Quintana Roo, I called the store, told the clerk that I was from the botanical garden, and ordered ten thousand springs. There was a long pause on the other end of the phone. "Uh, what are you going to do with all these springs?" he asked. I replied that I was working with a group of

Maya foresters in Mexico and that we were going to use the springs to make dendrometer bands to measure the growth of their mahogany trees. I explained that we planned to leave the springs out in the forest, which is why we needed stainless steel. Another long pause while he processed this information. Then, "I think we can give you a little discount on these."

Their return to their growth bands after one year represented a seminal moment for most of the ejido foresters. Although they had invested the time and energy to locate the sample trees and make the bands, there was clearly a nagging doubt in many people's minds as to whether the bands would actually work. I appreciate now that a lot of the work is done on faith—from both sides—during the early phases of community collaborations. The ejidos' doubt was put to rest after they remeasured the first sample tree. The band affixed around this individual, a large mahogany tree, had indeed expanded, and the caliper reading, taken several times, revealed that the tree had grown exactly 0.85 centimeters in diameter over the past year. There was a murmur of approval by everyone in the field crew. The next tree, another mahogany of similar diameter, provided additional insights into the growth of these trees. Caliper measurements, also taken several times, showed that this tree had grown only 0.18 centimeters in diameter. Given that the faster a tree grows, the more wood volume it produces, and the more money it makes for the ejido, everyone was curious about why one mahogany tree had grown more than four times faster than another during the same time period. A frenzy of chatter arose as the crew started assessing the canopy cover, the soil conditions, and the number of competitors around each sample tree. This is exactly what professional foresters do.

I went to the forest numerous times with the field crews to look at their sample trees, read the bands, and record growth data. I was impressed by how deep into the forest they had gone to set up their growth studies, and by the long distances between different clumps of sample trees. Most of the crews had GPS devices, but they did not use them to relocate their growth trees. They just knew where they were. I would follow the crews through the forest, walking for ten or fifteen minutes along an invisible—at least to my eyes—trail, and we would suddenly come on a group of trees with shiny bands and springs. We would read the bands, and then set off walking for another ten or fifteen minutes along another invisible trail to find the next group of trees.

In one ejido, the growth trees were located so far inside the forest that we elected to go by a three-wheeled, balloon-tired motorcycle. I asked, several times, in advance whether there would be room for me on the vehicle, and was assured that there was, that it would be "no problem"—they did this all the time. Four of us ended up piling on the utility trike. I climbed on last, and noticed a little sign (in English) on one of the back fenders—the one that I would be sitting on—warning in bold letters that the maximum capacity of the vehicle was one driver and less than ten kilograms of field equipment. We raced along a muddy logging road for about half an hour, skidding and swerving, and I almost fell off several times. We found all the sample trees, read the bands, and, in good spirits, headed back to the village. No problem.

Walk through forests in the Selva Maya and you will pass a lot of little hills. Some of these hills are hard patches of limestone that have resisted erosion while the surrounding area was gradually washed away and flattened. Geological processes alone, however, have not produced all the topography in the forest. The first evidence of this is a prevalence of stones with flat sides. Keep looking, and the stones will appear to be fitted tightly together—even, in a few places, to make steps. The whole hill is covered with soil and trees, but the slope is atypically constant, smooth and steep. Then you will come on a section of the hill where the stones have collapsed, and a small cave or chamber is visible within. If you ask the field crew, they will affirm that the hill is actually a Maya pyramid and add that there are *dozens* of them scattered throughout the forest.

During the Caste War of Yucatán (1847–1901), the Icaiche Maya formed a walled settlement in the jungles of Quintana Roo called Chichanhá. The war, a revolt by the indigenous Maya (primarily Icaiche and the Ixcanha) against an oppressive local government comprised mostly of Spanish officials, created an independent Maya state, Chan Santa Cruz, and drove all nonnative people from the region. I had never heard of it, but on one of our trips to the forest to check growth bands, we visited the ruins of the settlement with a young archaeologist from Veracruz who told us the story.

Rather than negotiate with the Spanish during the final stages of the war, the Icaiche rebels moved deep into the forest and created the walled city, which included schools, markets, public buildings, a church, and a deep cistern to provide water for the community. Access to the settlement was tightly controlled. Other Maya groups, as well as the Mexican government, made repeated requests for a meeting to investigate the possibilities of a cease-fire, but the Icaiche consistently refused such offers. As a final resort, the Catholic Church offered to send the local bishop to Chichanhá to say mass. The rebel leaders agreed, the bishop came and left, and, within a few months, all of the residents of Chichanhá had contracted a fever and died.

It took us a long time to walk to Chichanhá, and the absence of a clear trail suggested that villagers rarely came to visit. We saw huge *chicle* trees (*Manilkara zapota*), heard the occasional screams of howler monkeys (*Alouatta pigra*),[3] and touched the crumbled walls of an indigenous rebellion. Nobody said much as we walked through the ruins. We tossed a couple of rocks into the cistern to see how deep it was, but we never heard them hit the bottom. The

Selva Maya had grown back and erased much of the evidence of the once thriving Icaiche settlement. I reflected on the intensive farming and forest exploitation that must have been practiced by the residents of Chichanhá on this same site a hundred years ago, yet I saw little to suggest that any of these management activities had ever occurred. The forest seemed pristine.

At 2:37 a.m. on August 21, 2007, Hurricane Dean, a category 5 storm with winds of up to 265 kilometers per hour, slammed into the Yucatán Peninsula. The center of the storm passed right through the middle of the Selva Maya, destroying over one million hectares of forest and toppling or decapitating a large number of the sample trees we had banded. The six hundred or so bands that were still in place provided invaluable data on the growth response of tropical trees to hurricanes. Such information was certainly worthwhile, but not what we had in mind when we started the research.

All the ejido forests in our growth study were affected by the hurricane, but to different degrees. The forests that were hit the hardest were, for some reason, those with the highest densities of mahogany. These ejidos consistently made the most money from their forestry operations, and they were also the communities that had the necessary equipment and manpower to clean up the mess. The problem was what to do with all the salvage timber. If it was dumped on the market, the price, even for certified mahogany timber, would plummet. The other ejidos, whose forests contained low volumes of mahogany that consistently generated smaller revenues, escaped the gale-force winds. These ejidos never had the highest-value timber, but in the years following the hurricane they did have *some* timber to sell. In many respects, Hurricane Dean leveled the playing field in the forestry sector of the Selva Maya.

Hurricanes are a relatively common occurrence in Quintana Roo, and what invariably happens during the dry season in the months following one is that all the fallen trees and slash in the forest catch on fire. Even communities that proactively do salvage cuts and clear the slash from the understory run this risk. All it takes is one neighboring ejido that cannot afford to clean up its forest. In an attempt to avoid this scenario, representatives from several forestry ejidos in the Selva Maya went to Mexico City following Hurricane Dean and petitioned the government for emergency funds to help with the cleanup. The situation became complicated, all the work had to be put out for bids, the funds were delayed, and while all of this was going on, the forests caught on fire in Quintana Roo. We lost a significant number of growth bands in these fires.

In spite of disruptions by hurricanes and wildfires, my collaboration with ejidos in the Selva Maya produced several results of value. Local foresters now have a tool to precisely measure the growth of their timber trees, as well as a deeper understanding of the myriad environmental factors that influence tree growth and yield. The forest management plans produced by the ejidos are also more reliable—and sustainable—because the growth rates used to calculate the annual allowable cut were collected from sample trees in their harvest areas. All the communities involved in the growth study have greatly improved their chances of obtaining or renewing forest certification. The result, however, that made the strongest impression on me was that local communities, with little encouragement, quickly grasped the importance of measuring tree growth and were motivated to band and monitor the annual diameter increment of thousands of timber trees. Few forestry operations in the tropics invest the time and money to measure tree growth; fewer still incorporate these data into management plans. The ejidos, with the help of modern science, are doing both.

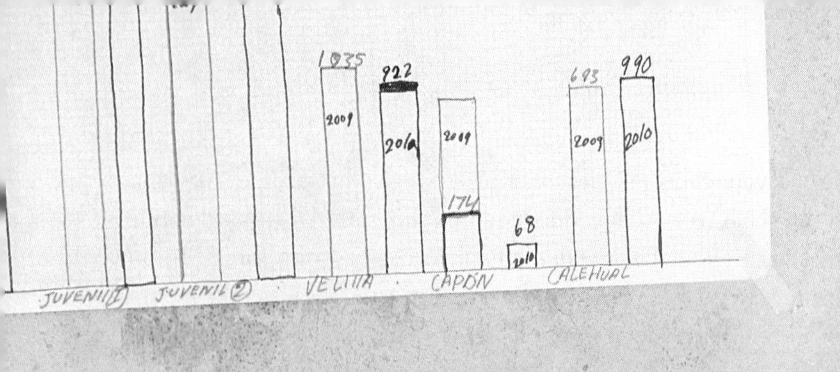
1035
2009
922
2010
2009
174
68
2010
693
2009
990
2010
JUVENIL(1)
JUVENIL(2)
CAPÓN
CALEHUAL

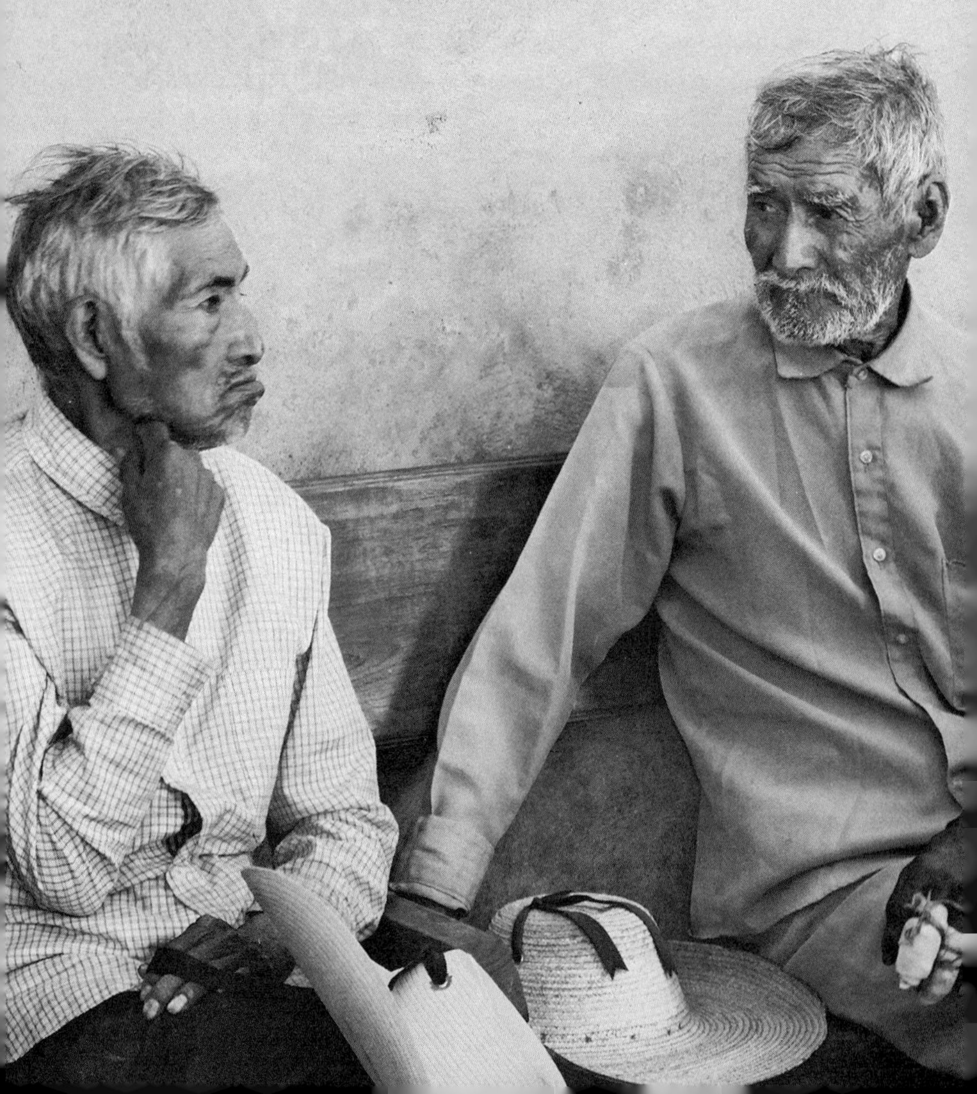

TWELVE

Managing Agave, Distilling Mescal

Acateyahualco (N17°46′11″, W99°1′34″), Municipality of Chilapa, Guerrero, Mexico, 2007–2013

Rural communities in the dry regions of Mexico harvest the leaf bases of several species of agave (Asparagaceae) plants to make mescal, a fermented, distilled, strongly alcoholic beverage. The beverage is traditionally served at religious celebrations, weddings, and other cultural events throughout the country, and large quantities are produced and sold each year. Harvesting commercial quantities of wild agave from tropical dry forests to make mescal might seem counterproductive since every species of agave used to make mescal flowers and fruits only once, then dies.[1] To ensure that the leaf bases accumulate the highest concentration of sugar for fermenta-

(*Opposite*) Master *mescaleros* from the village of Acateyahualco in Guerrero, Mexico. Note the histogram showing the size structure of the *Agave cupreata* population taped on the wall behind them.

tion, the harvesters chop off the flower stalk before it has a chance to develop. The leaf bases will be saturated with sugar when they are dug up, but the plant never has a chance to make seeds or to regenerate itself in the forest. If all the flowering agaves in a given area were harvested each year to make mescal, before long the population in that area would disappear. Further threatening the crop, tropical dry forests are the most endangered forest ecosystems in Mexico, even more at risk than tropical rain forests.[2] Once all of their agaves have been harvested, communities will have little incentive to maintain the forest instead of converting it to a cornfield or a cattle pasture.

The international market for mescal has grown rapidly during the past decade, and the commercial success of tequila, which is also made from agave, has prompted several large distilleries to start producing their own mescal.[3] In contrast to community-level modes of mescal production that are based on the harvest of wild agave in dry forests, large-scale commercial efforts are focused on the establishment of extensive agave plantations. To fortify their control over the mescal market, the large distilleries have lobbied the Mexican government to restrict community production of mescal by claiming that wild populations of agave are being dangerously overexploited.

I made my first trip Chilapa, Guerrero, in 2006 at the invitation of a Mexican colleague to visit a few *mescalero* communities and assess the regeneration status of local agave populations. What I found, which did not surprise me, was that the communities were well aware of how to harvest their wild agave species, known locally as *maguey papalote*,[4] without overexploiting it. A select number of adult plants were left unharvested each year and allowed to set seed. The seeds were collected, dried, and then broadcast throughout the harvest area. When I walked through a tract of dry forest being exploited for mescal by the community of Acateyahualco, I saw hundreds of thousands of agave seedlings growing in the rocky soil. Rather than posing a threat to an endangered forest ecosystem or causing the depletion of wild stocks of an important resource, the small-scale production of mescal by communities provides a powerful lever for conservation.

Prohibiting communities from continuing to produce mescal locally would be a cultural and economic disaster. Culturally, mescal is a fundamental component of many aspects of Mexican life. Economically, it provides local communities with an important source of income. And from an ecological standpoint, it would remove the only incentive that most rural communities have to conserve the few remaining tracts of dry forest on their land. Fascinated by what I saw in Acateyahualco, I started a long-term project in 2007 to document how the local communities were managing their agave.

Our first meeting at Acateyahualco, called so we could explain what we were doing and ask for assistance from the community, was attended by everyone in town. The room was filled with men in huaraches and cowboy hats, women with long braids and beautiful, embroidered blouses, and a few babies. Several dogs wandered in and out. On a bench in front sat three wizened, older gentlemen with gray hair and scraggly yet distinguished beards; all wore well-washed, long-sleeved dress shirts with the top button buttoned. I did not need to be told that these were the elder mescaleros of the community, men who had been making mescal from the wild agave plants growing in local forests for forty or fifty years. This is what traditional knowledge looks like, and these were the three men we needed to convince of the utility of resource inventories, quantitative data, and histograms.

We gave our presentation, and everyone listened intently. At the end we had a lot of questions—from both the men and the women—and good discussions. The three elders sitting up front would periodically put their heads together and whisper to one another. By the end of the meeting, the villagers agreed that it would be more useful to find out exactly how many agave plants there were in the forest and what size they were than continue harvesting the resource based on a general impression that there were "still a lot of agaves out there." We agreed to show them how to inventory the resource and produce a management plan if they would agree to collect the data and do the periodic monitoring. Hands were shaken, after which a wheelbarrow containing several cases of half-liter bottles of Coca-Cola and packets of cookies was rolled into the room. Little plastic cups were also passed around for the obligatory *copita* (or two) of Acateyahualco's finest mescal to toast the new collaboration.

Although the basic concept is the same, counting and measuring agave plants is quite different from doing an inventory of trees. We do not measure diameter, or diameter growth, or merchantable height, or even canopy cover. Rather, we count the plants and put them in size classes, and then observe the rate at which they die or move into the next size class. The procedure is easy, but it is important to select size classes that everyone counting can recognize. After discussions with the community, we divided the agaves into five size classes—seedlings, three juvenile classes, and adults—and made a sixth class for harvested individuals. Seedlings were tiny plants with fewer than five leaves, while the juvenile classes were based on height. To make things as easy as possible in the field, we separated the juvenile classes so that they could be measured by the mescaleros using their leg as a ruler—class I juveniles were no higher than the ankle, class II juveniles came up to the knee, and class III juveniles came up to the waist. Plants classified as adults displayed the first signs of an inflorescence, and the final class, harvested, was reserved for individuals whose inflorescence had been cut and the leaf bases, or *piña*, extracted.

The system worked perfectly, and in spite of the obvious height differences among villagers, all the agaves were assigned to the appropriate size class. The only field equipment needed was a compass, a twenty-meter-long rope, and a clipboard for recording the data. It took a couple of days to persuade the field crews to leave their rifles behind during the agave inventory. I think they finally stopped bringing their guns because we were making too much noise and scaring off the animals.

One of the older mescaleros from Acateyahualco joined a field crew a couple of days after the inventory started. Dressed in a classic, sweat-stained sombrero, a dusty pair of huaraches, and old khaki pants cinched high above the waist, and wielding a large machete, this man had been harvesting agave and making mescal most of his life. Counting agave plants, however, had never been part of the process. But after our meetings and discussions in the evenings with people who had been out in the forest all day collecting data, he felt that running transects was probably something that he should know how to do.

Experienced mescal producers in Acateyahualco gauge the quality of their product by pouring the newly distilled mescal into a calabash (gourd) cup and examining the density of bubbles that form on the surface.[5] I asked whether the density was a measure of alcohol content and was told that it was not, but more bubbles indicated a more flavorful mescal. I was also told that commercially produced mescal, made from agaves grown on plantations, produces *no* bubbles. Despite repeated inquiries, I was unable able to find out why wild agave produces a bubblier mescal than plantation stock.

During the original meetings at Acateyahualco, I suggested that a five-year monitoring cycle would be sufficient for the villagers to detect changes in the size structure and regeneration rate of the agave population. After conducting the first inventory, the community could then wait five years before having to repeat the fieldwork. Since it had taken five crews eight days to collect the first set of data, and the villagers all had a lot of important things to do besides running transects, I figured that this would make the program a bit more manageable. But the community did not want to wait. They argued that much could happen in five years, that they were fully committed to the idea of sustainably managing the agave on their land, and that they were going to repeat the dry forest inventory *every* year. I was overwhelmed by their commitment, but I suggested that if they found the task too difficult or time consuming, one inventory every couple of years would provide enough data to assess the impact of their agave harvesting. We had a final little plastic cup of mescal, and toasted one another—and I left for five years.

I returned to Acateyahualco in May 2013 to see how things were going. The walls of the meeting room were covered with dozens of multicolored graphs and histograms showing the number of agave individuals in each of the six size classes, and once again, everyone in the village was there. The village head convened the meeting, after which selected leaders of the mescal group presented the results of the agave project. They had, in fact, repeated the inventory of the agave populations every year for five years. What had taken the field crews eight days to do the first time, they were now able to complete in *one* day. They were becoming faster and more efficient in the way they collected these data, with the same teams working the same areas each year. All the original tally sheets were neatly clipped into folders marked with the year.

They went through the histograms showing the inventory results year by year. The community had continued to commercially exploit agave in the dry forest each year, and the number of agave adults harvested had increased slightly over the five-year period. The number of individuals in the juvenile classes bounced around considerably from year to year, but the smaller classes seemed to be getting larger. The final histogram showed the number of harvestable adults in the population for each of the five years. Although this figure also fluctuated a bit, the overall trend across the five data points was flat. The number of adult agave plants growing in the dry forests at Acateyahualco had remained essentially constant through five years of commercial resource exploitation—and the mescal team had quantified this through their inventories. So much for the allegation that the community was overexploiting its wild agave. This time, I was the one who asked that the little plastic cups of mescal be passed around, and I offered a heartfelt toast to the diligence and deep sense of stewardship of the mescal producers in Acateyahualco. I was, and still am, very impressed by what they were able to accomplish.

The situation in Guerrero worsened after my final trip. Drug violence escalated, the teachers went on strike in response to proposed educational reforms, and the central government, as part of its National Crusade Against Hunger, sent the army into rural villages to prepare food.[6] Currently, there are roadblocks on many roads, and trucks loaded with soldiers in riot gear circulate around the state. Heavily armed masked men loyal to a local drug gang have taken over Chilapa, and in September 2014 forty-three students were killed outside Iguala. Successful examples of community-based resource management lose their newsworthiness when all this is going on. The community of Acateyahualco has been exploiting the wild agave plants in their dry forests to produce mescal for several decades. They have controlled the rate at which adult plants are harvested each year to ensure that local populations continue to regenerate, and they have conducted systematic inventories and annual monitoring surveys to quantify the sustainability of their harvest activities. Five years of data have shown that the traditional method of producing mescal in Acateyahualco *is*, in fact, sustainable. This should be front page news.

THIRTEEN

Landscape Dynamics in Southwestern China

Three Miao communities in Leishan County, Guizhou Province, China, 2010–2011

The forests of Guizhou Province have been used, managed, and conserved by traditional Miao people and their ancestors for over a thousand years. The Miao are one of the fifty-five recognized ethnic groups in the People's Republic of China; *Miao* is the collective noun used to describe this group of linguistically related ethnic minorities, some of which are Hmong. These have not been easy times for either the Miao or the forest. There have been repeated migrations, resettlements, civil unrest, intermittent rebellions against the Han Chinese,[1] malaria outbreaks, and periodic burning of the forest—and Miao villages—by government armies trying to drive them out of the region. Through it all, the Miao have tended their rice fields, harvested timber to build their houses, collected forest fruits and me-

(*Opposite*) Miao children in their finery at Langde village, Qiandongnan Autonomous Prefecture, Guizhou, China.

dicinal plants, and consistently enriched local forests by planting useful tree species. These low-level, long-term silvicultural treatments have had a major impact on the forests of Guizhou.

Traditional patterns of Miao resource use in Guizhou, however, have been changing since the founding of the People's Republic of China in 1949, in response to shifting government policies, population growth, and decreased access to forest resources. Although forests still provide an important source of building materials for house construction, fruits, fibers, resins, mushrooms, medicinal plants and other non-timber forest resources are being exploited and managed with decreasing frequency. The diminishing importance of these subsistence plant products has been accompanied by a reduction in the diversity of useful taxa in the forests.

Government incentives and planting programs have replaced the normal successional development, or fallow phase, of Miao agricultural fields with single-species plantations of *shamu* (*Cunninghamia lanceolata*), a particularly valuable local timber tree frequently referred to, erroneously, as Chinese fir.[2] The total area of forest in many Miao regions has increased as a result of this planting activity, an achievement much publicized by the State Forestry Administration, but the variety of plant communities within the local landscape, as well as the taxonomic diversity of the species found in them, appears to be decreasing.

The Miao region of southeastern Guizhou is one of the most important and expansive forest areas in China. But not all forests are alike. The mosaic of species-rich forest vegetation originally created and maintained by the Miao is being homogenized by forest policies that focus on timber production and promote a higher degree of state and private-sector, rather than community, control over forestland. In 2010, I started collaborating with Minzu University, in Beijing, a school for non-Han Chinese, and we developed a project to document the changes in forest structure and composition that were occurring near Miao communities in the Qiandongnan Autonomous Prefecture, in the southeastern part of Guizhou Province.

The work in Guizhou was focused on three Miao villages that varied in size and in distance to Leigongshan Nature Preserve, a protected area. Larger villages, we speculated, would put a higher demand on the forest, while communities located closer to the protected area would have access to larger tracts of forest but would also be subject to more stringent regulations and enforcement. The composition and structure of the forests within the protected area were of special interest, given that Miao communities had been using and grooming them for hundreds of years before the land tenure was changed. These forests would probably provide the clearest examples of what the original vegetation of the region looked like.

The basic workflow in each village was the same. We first conducted a series of unstructured household interviews to compile a list of important forest resources and estimate the quantity of each resource that was used in a year. The objective of these household surveys was to come up with a preliminary assessment of the current *demand* for forest resources in each village. We then went to the forests—visiting examples of all the different forest tenures—to quantify the *supply* of these resources. The estimated demand was then compared to the observed supply to enable us to characterize the trajectory of the interaction between people and forest in each place. This would help us speculate about future forest use for the Miao villagers.

The largest Miao village in the survey contained more than thirteen hundred families and was a beautiful, expansive cluster of traditional tile-roofed wooden houses clinging to the steep slopes along the banks of a river. The forests were located in the mountains behind the village; all the flatter land was in terraced rice fields. The setting was gorgeous, Miao women wearing exquisite silverwork and detailed embroidery were casually strolling around, and the large plazas and cobbled sidewalks were teeming with Chinese tourists.

The second village had about a hundred Miao families and was located in the buffer zone of the protected area. The houses were also made of wood, with the same kind of roof tiles, and they were equally well crafted, but they appeared to be more worn and lived in; the rice fields were less manicured. No one was walking around the village in traditional Miao attire. The village had been in its current location for 350 years.

The last Miao village, about two hundred families, was located inside the core zone of the nature preserve, and as a foreigner I was not allowed to enter. But we obtained permits so the Chinese researchers and students collaborating with the project could visit the village to do the household surveys and forest inventories. During the second year of the project, the boundary of the core zone of the reserve was moved to exclude this village. The village was not forced to move out of the core zone; the State Forestry Administration simply moved the boundary. I would speculate that the pattern and intensity of forest use by the village did not change in the slightest in response to this cartographic modification.

Once we started going down the side streets of the largest, most touristy village, it no longer felt like Disneyland. Our household surveys revealed that four small villages were merged in 2006 to form the current village; originally there had been houses on the site where the main plaza was constructed. The local authorities had expropriated the land and forced the residents to move; most of the people said that they liked their new house, though not their new location, better. They complained about the outsiders who had moved in and opened restaurants and hotels. All the households interviewed still walked several kilometers to the forest each week to collect firewood and harvest mushrooms and chestnuts.

As recently as thirty years ago, all the forestlands outside of the village were community forests. A village council, rather than a district office, decided how many trees could be cut and who could cut them, which species needed to be planted, and where would be the best place to plant them. These days, a permit from the government is required to cut or plant anything.

We walked about two hours up the mountain behind the village to run an inventory transect in a tract of community forest. We saw a few big shamu trees, but the canopy was mostly oaks, sweet gums, chestnuts, and beeches. The understory was filled with rhododendrons, hydrangeas, and the occasional tea plant. We saw many different tree species, few of which I knew, and none of them was in flower—it was a good thing we had a botanist with us. Yang Chenghua from the Guizhou Forestry Academy not only wrote the scientific names of all of the trees in his field book after each transect, he also made spreadsheets with the names of all the associated trees, shrubs, and herbs that he recorded on each site. In several places, the transect line crossed a shallow, five centimeters deep by sixty centimeters wide drainage channel. The channel had been compacted to the hardness of a ceramic tile and seemed to be ancient. The locals told us that a network of these channels ran throughout the forest; they had been built centuries ago to collect water from the forest and route it down the mountain to the rice fields. They were still in use.

We were exhausted when we finished the transect and started walking down the mountain. One of the foresters on the team had called back to town to get a truck, and it was waiting on the road for us when walked out of the forest. We all piled in—which was a bit tight, because there were ten of us—and slowly bumped and swerved our way back to town, arriving in plenty of time to bathe and take our place on the main plaza with all the Chinese tourists to watch the late-afternoon Miao dance extravaganza.

The forests surrounding Miao villages are currently subject to three different types of tenure. State forests are those in which both the ownership and use rights belong to the central government; community forests are the collective property of local communities that control the ownership and use rights; and private forests, which are also collective property, are those in which individual households have the ownership and use rights of trees and other forest resources. The various forests have different names and caretakers, but all were created from the same original forest base that the Miao have been farming, exploiting, and managing for hundreds of years. Our inventory transects revealed that there were few differences between what was growing in the state forests and the community forests.[3] Private forests, on the other hand, were a different case. Individual households have created these short-rotation plant communities to produce timber that can be sold or used locally for building. These forests contain shamu, a single species of pine (*Pinus massoniana*),[4] and not much else. They appear as green spreads on a satellite photo, they protect the soil to a certain degree, and they provide a source of revenue to selected households. But from a floristic, structural, or functional standpoint, they bear little resemblance to the forests the Miao created.

We detected an atypical abundance of shamu trees in both community and state forests. This endemic conifer is one of the most important timber species in China, and Miao villagers preferentially use the wood to build their houses and to make furniture, coffins, and agricultural tools. The species, however, characteristically occurs as scattered individuals in mixed deciduous and evergreen broadleaved forests, not in the densities of 500–600 individuals per hectare that we recorded in our transects. Given the extreme value and utility of the species, the long-term proximity of shamu populations to human settlements, and the demonstrated management capabilities of the Miao, it seems clear that these forests, even the ones currently under state control, have been purposely enriched with a useful species.

Miao villages are allocated a selected number of shamu trees each year, and this timber is distributed among households based on a permit system controlled by the provincial forestry department.

Some fifty to sixty shamu trees are needed to build a typical Miao house, and it may take a decade of annual permit applications for a householder to get enough wood to finish it. Several people told us that the permitting process had taken less time for them because their original house had burned down and they were given preference when the shamu logs were allocated. Other landholders confided that if the shamu trees in their private forest were burned in a fire, they were allowed to do a salvage cut to save the timber. Fire presents a continual risk to people living in tight clusters of old wooden houses, and many of the larger villages boasted a network of fire hydrants scattered around town. The municipal authorities and village leaders assured me that everyone was extremely vigilant and fires never occurred in the region. Nonetheless, a small, carefully controlled burn appears to significantly reduce the time required to get the wood needed to build a house.

Once a family has a stack of shamu logs and a place to build, the next step is to turn the logs into boards. Not everyone has the means to do this. I met a fellow during one of the household surveys who had built a portable sawmill that householders could lease, and he would set up his mill behind the house under construction and make boards. He moved from village to village, going wherever he was needed, and said that he helps build about three houses a year.

The Miao make their houses out of shamu poles and planks using simple post-and-beam construction. The two structural components are held together with a mortise-and-tenon joint locked with a wooden peg, a centuries-old technique that is still in use. The boards used for the walls, windows, and doors are carefully countersunk into grooved baseboards. All the boards are neatly planed and the joints carefully chiseled and tightly fitted together using a mallet. No nails are used at any point in the construction process. I have been in Miao houses that are three stories tall and more than thirty years old.

The Grain for Green program, also known as the Sloping Land

Conversion Program, was initiated in the late 1990s as a way to control soil erosion in areas of mountainous terrain. Farmers who had agricultural fields on lands with slopes greater than 25 percent could take these fields out of production and plant them to trees, and the Grain for Green program would give them sacks of rice and a cash subsidy every year for doing so. Many Miao farmers took advantage of this program, and the steep slopes surrounding their villages are covered with young Grain for Green forests. In terms of scale, budget, and duration, it is one of the largest payment-for-ecosystem-service experiments ever attempted.

Before the revolution, the Miao would also periodically take their agricultural fields out of production when yields started to drop. The site would be allowed to go through a natural successional process back to forest, and the fallow regrowth would be enriched with several species of fruit trees and shamu. After several decades, the forest would contain dozens of tree species and a variety of useful forest resources. Once the timber trees in the forest had attained a merchantable size, they would be harvested, and the site would be cleared and recycled back into agriculture. The Miao landscape was a mosaic of young forests, old forests, and agricultural fields, each vegetation type continually growing into or being replaced by another type, based on the needs and collective decisions of the village. I was told that soil erosion had never been a problem when this system was in place.

The Grain for Green program interrupts this indigenous system of land management by introducing a new vegetation type that can never become anything but another rotation of itself. Farmers are provided with planting material for the program, and based on our household surveys the choices are limited to shamu, one species of pine, and tea. Farmers are not allowed to enrich these plantings with other species, and the plantations are periodically checked to make sure the farmers are complying and that the shamu, pine, and tea trees are being well cared for. Additionally, the farmers do not own the shamu or pine trees they have planted. These trees, when they attain a commercial size, will be allocated through the existing permit system just as all the other timber species are—unless, of course, there is a fire.

Every year fewer Miao families are involved in the management of community forests. Decisions made in the village council currently are focused more on allocating shamu logs for construction than on planting fruit trees, scheduling a thinning operation, or maintaining the drainage channels that supply water to rice fields. Existing forest policies provide little incentive for the Miao to use the formidable silvicultural skills they have gained from living in the forests of Guizhou for millennia, and the original species-rich forests of the region are being slowly converted to monocultures.

We visited a well-known Miao village in Guizhou that was founded during the Ming Dynasty; the residents had been living and farming in the same place for more than six hundred years. The villagers were busy planting their rice fields when we arrived, and we were able to wander quietly along the trails and through the back alleys of the village without being disruptive. Both the men and women wore work clothes with their pants rolled up, and were digging, transplanting, weeding, and periodically tapping the water buffalo with a stick to make it get up out of the mud and leave the rice field. (I had never before realized that farmers used a string to ensure that the rice seedlings are planted exactly in straight lines.) At the sound of a loud gong in the village, they all finished what they were doing and headed back to town.

The men put on dark pants and jackets, and the women changed into their embroidered skirts and blouses, silver necklaces, earrings, bracelets, and elaborate headdresses. The transformation was magical. A busload of tourists had just arrived, and the Miao were preparing to offer them a welcome dance. We stayed to watch the performance, which was lovely and much appreciated, but I was glad that I had had the opportunity to peek behind the curtain before the bus arrived.

FOURTEEN

The World of Rattan

Communities and forests within the Greater Mekong Region of Laos, Cambodia, and Vietnam; six nature reserves within the Truong Son Mountains of Vietnam, 2009–2014

Rattans are spiny, climbing palms that grow throughout the Old World (Africa, Asia, and Australia) tropics. There are currently over five hundred different species of rattan known, and the number continues to grow as more botanical collections are made. Rattan palms are used for a variety of subsistence purposes—cordage, basketry, food, medicine, and thatch—and the flexible stems, or canes, provide the raw material for a multi-billion-dollar international furniture industry. Almost all of this material is harvested from *wild* populations. Several million people use, collect, and sell rattan, or are involved in some way with the international rattan trade. Rattan

(*Opposite*) Rattan (*Daemonorops jenkinsiana*) stem with fruits, Central Truong Son Mountains, Vietnam.

is, without question, one of the most important non-timber forest products in the world.

Over the past fifteen years, I have studied rattan in a number of different countries and contexts.[1] Much of this work was done in collaboration with a palm systematist, Andrew Henderson, who was extremely skillful at identifying both named species and new species when we happened to find them. In each of these projects we have been confronted by a similar set of difficult conditions.

Local names for rattans vary considerably, and the exact taxonomic identity of many commercial species is still in doubt. It is discouraging not to know the Latin name of a rattan that generates millions of dollars in foreign exchange to a country each year. In addition, almost no information is available about the density and conservation status of wild rattan populations. We have been told that "rattan stocks are dwindling," but we know virtually nothing about the population densities of wild rattan, or whether these populations are regenerating, or which species are most resilient—or most susceptible—to the impacts of repeated harvesting. Finally, most of the rattan cane is harvested in an uncontrolled manner. Harvest quotas, when they exist, are based more on the current demand for the resource than on the supply of cane; consciously managing wild rattan populations, although occasionally discussed, is rarely implemented.

Andrew Henderson and I first started to engage directly with the management of wild rattan populations in six protected areas in the Central Truong Son Mountains of Vietnam. Vietnam is the third-largest producer of rattan cane in the world—about one out of every six canes sold worldwide comes from Vietnam—but much of this material is collected illegally from local nature reserves. Our work in Vietnam subsequently expanded into Laos and Cambodia, where rattan is harvested commercially from community forests. The rattan species and forest habitats we encountered were similar throughout the region, whereas the accessibility of the rattan resource, the expertise of local collecting groups, and the motivation to manage rattan on a sustainable basis varied from country to country. By the end of our project, hundreds of communities in the Greater Mekong had invested in managing their rattan, and a well-known international furniture retailer was buying the annual harvest.

Much of the rattan coming out of Vietnam is collected from a network of protected areas in the Central Truong Son Mountains. This region contains the highest density of commercially valuable rattan species in the country—as well as the highest concentration of rattan collectors. The terrain in these protected areas is rugged and mountainous, traversed by the Ho Chi Minh trail (which is actually a network of trails linking Vietnam, Cambodia, and Laos that was used as a major supply route by the Viet Cong during the Vietnam War). Fifty years ago, American and South Vietnamese forces relentlessly bombed and napalmed these forests to interrupt the flow of supplies along the trail. Rattan, however, adapts well to disturbance (to put it mildly) and high light levels, such as were created by the bombs' thinning of the trees, and today the forests of the Central Truong Son produce several million dollars' worth of rattan cane each year.

But at the time we began our project, rattan collectors were not supposed to be going into the nature reserves, especially the core zones of the nature reserves, to collect rattan. Their doing so represented a dilemma for the local Forestry Department. On one hand, rattan collecting generated sizable revenues and offered employment to thousands of villagers. On the other, the creation of these nature reserves was facilitated by generous donations from several international foundations and conservation organizations, and these groups had started to *strongly* request that rattan collectors be kept out of the reserve. My contribution was the suggestion that rattan, like any other forest resource, could be managed in such a way that it would not be depleted. While the rattan in the core zone should probably still be protected, communities *could* be allowed to harvest it in the buffer zone if they did so sustainably. It was suggested that there might be some legal precedent in the Forestry Law to allow communities to do this, but no one had ever made the attempt.

My initial questions concerning the species and quantities and growth rates of rattan found in the nature reserves went unanswered. At that time, the situation of the rattan in the Central Truong Son Mountains could best be characterized as an extreme example of a valuable forest resource of questionable taxonomic identity being exploited in an uncontrolled fashion until supplies were completely depleted. We contacted the directors of six nature reserves—Bac Huong Hoa and Dakrong in Quang Tri Province; Phong Dien and Sao La in Thua Thien Hue Province; Song Thanh in Quang Nam Province; and Ngoc Linh in Kon Tum Province—to put a project together.

The aim of the project was to train the forestry staff at each reserve to identify the local rattans and run a simple inventory transect. We would then provide financial support to enable two field crews in each reserve to identify, count, and measure wild rattan plants. By analyzing these data, we could learn which rattan species grew in each reserve, at what densities they occurred, and whether their populations were regenerating in the forest. We could then select the best species to manage, initiate growth studies, make a few yield predictions, and start talking to communities about *legally* harvesting rattan in the reserve.

Representatives from all six nature reserves attended a three-day planning workshop. We provided the participants with a draft field guide in Vietnamese, with pictures of the leaves, stems, and reproductive structures of each of the rattan species that we thought they would encounter in the inventories. The pages were laminated so the guides could be used in the field. If crews found rattans that were not listed in the field guide, they were instructed to make a collection and take copious notes on the plant. By having teams from different reserves looking at the same species, we hoped to come up with a more or less standardized local nomenclature. There are over thirty different common names for rattan in Vietnam, and the same name is used to refer to different species in different places. We did our best to ensure that our teams called the plants by the same name.

We explained the inventory methodology, showed everyone how to use the equipment—a twenty-meter transect rope with knots to correct for slope, a compass, a clinometer for measuring slope and rattan heights, and a GPS device—and spent several days in the forest running inventory transects. We experienced days when we tallied only a few rattan species, days when we tallied a dozen, days when the topography was so steep that we had to stop because it was too dangerous, and one day when everyone got covered in leeches. We spent the last day of the workshop devising a system for locating the transects that would ensure that every part of the reserve was sampled, instead of all the transects running a short distance down the road from the park headquarters. We divided each reserve into four concentric bands with the park headquarters in the center, and the field crews were asked to randomly locate a certain number of transects in each band. We reached a consensus about the number of transects, the per diem rate, and the time allocated to complete the fieldwork, and the teams were given a contract for the director to sign. Each reserve agreed to sample 160 transects and complete the work in six months. We had a delightful farewell dinner together, and I went back to Hanoi and got a plane to New York.

In spite of conscientious training, enthusiastic participants, and signed contracts, we can never be sure that everything in a project will work out as planned. We had a capable and well-connected local project coordinator and a forest resource worth millions of dollars. All the reserves had a strong incentive to complete the inventories —but even with these advantages things can go wrong. Eight months went by. Then one day our project coordinator showed up with fifteen orange field books filled with inventory data. In spite of rain and a continual series of operational difficulties, the rangers and field assistants at the six reserves had completed 960 transects and counted, measured, identified, and spatially referenced 175,295 rattan plants in the Central Truong Son Mountains, the largest rattan inventory ever conducted. Impressive as these statistics are, that the Truong Son rattan inventories merit this distinction goes a long way to explain the current pattern of resource depletion characterizing the international rattan trade. If less than two hundred hectares is the largest quantitative assessment ever made of a multibillion-dollar wild-harvested commodity, an enormous amount of work clearly needs to be done in this sector.

Discovering the names and quantities of the different rattans in a forest is only the first step in a sustainable-use equation. The number of harvestable canes that exist in the forest at a particular moment in time can be thought of as the *stock* of the rattan resource. But this rattan cannot all be harvested each year; harvesting the entire stock of a plant species is how resources become overexploited and ultimately depleted. Most important for the sustainable harvest of rattan is gauging the annual *yield* of the resource—

the quantity of new cane, or growth, that is produced each year by the stock. This quantity represents the limit of how much rattan can be harvested from the forest each year. Unless the managers understand the growth characteristics of a wild-harvested resource, they cannot put a program for sustainable harvesting in place, or set harvest quotas, or make financial projections, or do economic planning. When we started our project, no data were available on the growth of wild rattans in Vietnam.

Based on the inventory results, we started growth studies on the three most important commercial rattan species in five of the nature reserves. In each reserve, field crews selected three hundred individuals of varying size from each species, estimated the height of the rattan, and marked the last leaf and bud on the stem with bright orange spray paint. Spray painting the last leaf and bud on a ten-meter-tall spiny rattan plant can be extremely difficult, not to say prickly. The crews frequently constructed rustic ladders in the field if they were unable to climb a neighboring tree. Twelve months later, the sample plants were revisited, and the distance between the orange paint and the end of the new, green stem tissue that was produced was measured to estimate the rate of annual growth. Each reserve marked and followed the growth of 900 rattan plants, providing a total sample size of 4,500 individuals. We went from no growth data to a great deal of growth data for six commercial rattan species in one year.

We learned many useful things from the growth studies—for example, that growth rate is a function of size. Tall rattans grow two to three times faster than small rattans. We also learned that the growth rates for most species were well over a meter per year, and that some species were growing in excess of two meters per year. Perhaps one of the most important things we learned, however, especially in terms of promoting the collection of further data on rattan growth in other countries, is that a reasonable estimate of rattan growth can be obtained with a sample size of about 120 sample plants per species—less than half the 300 plants that we originally marked.[2] In terms of logistics, a small field crew could collect the growth data needed to define a sustainable harvest of a commercial rattan species in two days.

We held a final management-planning workshop using the inventory and growth data. Members of the technical staff from each reserve attended, as well as the directors of two of the reserves and the vice director of a third. All showed much interest in managing rattan, and with the presence of three directors, who collectively control the activities in over 175,000 hectares of forest in central Vietnam, everyone listened carefully, asked good questions, and took copious notes. Two of our take-home points seemed to resonate most strongly with the workshop participants, that it is impossible to sustainably harvest from the forest in one year more than is produced in that time period, and that carefully collected inventory and growth data—which they now had—could be combined to provide an estimate of the total annual growth of rattan in a forest. It was generally agreed that these principles could be usefully applied within the buffer zone habitats of protected areas in Vietnam.[3]

At the same time we were looking at the supply side of the rattan resource in Vietnam, a large international conservation organization had started a project in Laos and Cambodia that was focused on the demand issues—processing technologies, market development, supply chains—of the local rattan sector. I met the director of this project, Thibault Ledecq, currently the regional forest coordinator for the World Wildlife Fund's Greater Mekong Program, a delightful Belgian with more than a decade of experience in Laos. Ledecq described what his rattan project was doing, I told him what our project was doing, and we both quickly agreed that we should combine forces and start working together.

I went to Laos and Cambodia several times and gave training workshops to his staff about forest inventories and sustainable resource management, while Andrew Henderson made a few trips to collect rattans and work with local palm researchers. I made site visits to participating communities in Laos and Cambodia to review their field operations and answer questions, and began collaboration on a growth study that had been initiated the year before involving several commercial species.[4] During the workshops and site visits, we talked extensively about rattan taxonomy, forest inventories, yield studies, community management planning, and the biological bases of sustainable resource use. We discovered new species and adapted the data collection methodologies developed in the protected areas of Vietnam for use in community forests.[5] We helped the teams simplify and streamline their procedures and disseminated the protocols to villages interested in managing their rattan resources. Despite a few logistical problems and conceptual difficulties, by the end of 2013, forty villages in Laos, thirty villages in Vietnam, and twenty villages in Cambodia had started managing their rattan resources using inventory and growth data that they had collected themselves.

The world's largest furniture retailer, IKEA, located in Sweden, sells a lot of rattan furniture. In an effort to make its supply chain more sustainable, IKEA had been supporting the conservation organization's demand-side rattan project in Laos and Cambodia for several years. We thought the company might also be interested in supporting our new rattan collaboration, which was addressing both the demand and the supply side of the issue, which ideally should be considered together. Increasing the supply of a resource when there is no demand is bad business; increasing demand without first making sure of the supply leads to resource depletion. The company was receptive to the idea, and started funding the work. IKEA also agreed to purchase a specified quantity of sustainably produced rattan cane from villages participating with the project.

One of the major outputs of the collaboration was a field guide to the systematics, ecology, and community management of the sixty-five species of rattan that grow in the Greater Mekong Region.[6] This guide was one of the first of its kind. Field guides to selected groups of plants, monographs about plant ecology, and textbooks of forest management have been available for years, but rarely have the three disciplines been brought together in a single volume, especially one focused on a non-timber forest resource that offers simple management protocols.

During my last site visit to Laos, we designed a monitoring system for assessing harvest impacts and the long-term viability of forest management. Such is the basic workflow. The crews collect the baseline data, define a sustainable harvest level, harvest the resource, and then monitor what happens, adjusting harvest levels as necessary. In 2011, the rattan forests of four villages in the Bolikhamxay Province of Laos were certified sustainable.[7]

Our projects in Vietnam, Laos, and Cambodia have provided useful examples of how rattan can be exploited on a sustained-yield basis, both in community forests and the buffer zones of protected areas. Harvesting wild rattan based on quantitative inventories and growth data, according to the productive capacity of each population, is clearly a major improvement over uncontrolled exploitation, and we have demonstrated how, with a little training, these basic data can be collected by villagers or forest rangers. It is not too expensive or time-consuming to manage rattan stocks, and now that we know the names of all the local species and have reliable estimates of the annual growth of several commercial ones, we hope that more communities and forest wardens in the Greater Mekong Region will consider investing in the long-term benefits of sustainable resource use.

FIFTEEN

Community Forestry in Myanmar

Forests in the Hukaung Valley and near the villages of Nam Sabi (N25°21′42″, E95°20′33″) and Shinlonga (N26°31′13″, E96°37′34″) in northern Myanmar, 2009–present

I have always wanted to work in Myanmar. The country contains over half the remaining tropical forests in Southeast Asia and has an astonishing diversity of indigenous ethnic groups. Unfortunately, it was also, until quite recently, controlled by one of the most repressive military regimes in the world; the resident Nobel laureate, Daw Aung San Suu Kyi, who received the Nobel Peace Prize in 1991, had been under house arrest since 1989. The Wildlife Conservation Society (WCS) had been engaged in a tiger conservation project in northern Myanmar for several years, and in late 2003 I

(*Opposite*) Villagers from Nam Sabi running the baseline to establish a hundred-hectare village management area in the buffer zone of the Htamanthi Wildlife Sanctuary, Sagaing Region, Myanmar.

called a colleague there and inquired about the possibility of doing collaborative work, particularly on projects in Myanmar that might need botanical, ecological, or forestry advice. His answer opened new vistas of possibilities.

Tigers occur at low densities in the forest, and they are adept at avoiding wildlife biologists. To collect data on tiger population sizes and migration patterns, researchers generally must rely on camera traps, small, remotely activated cameras triggered by a trip line, motion sensor, or light beam. A large number of these traps are set up in the forest, and when a tiger hits the trip line, activates the motion sensor, or crosses the light beam, the camera snaps its picture. After a designated period of time, the researchers revisit the traps to collect the film or memory cards and put fresh rolls and cards in the cameras. When the researchers from the WCS developed the images from their camera traps, they found several shots of tigers—but they also discovered a number of pictures of rattan collectors. These rattan collectors were not only harvesting inside a protected area; they were also a threat to the tiger population. If they saw a tiger, they would undoubtedly shoot it because a dead tiger was worth more than they could make from a year of rattan collecting. My WCS colleague bombarded me with questions about rattan: How many species are there in Myanmar? Can wild populations be managed? Is rattan exploitation compatible with tiger conservation? Clearly it was time to consider a joint project.

The last treatment of the palms of Myanmar had been done in the late 1800s, and few rattan specimens from Myanmar could be found in any of the world's herbaria, while virtually nothing had been written about the local forests, the non-timber forest resources growing in them, or the indigenous communities that exploit or manage them. A lot of work needed to be done. The camera trap pictures had come from the Hukaung Valley Tiger Reserve, now the Hukaung Valley Wildlife Sanctuary (HVWS), in Kachin State, the largest protected area in Myanmar. Established in 2001, the HVWS extends over 6,300 square kilometers. I made my first trip to Myanmar in 2004, together with my palm botanist friend Andrew Henderson, to set up an expedition to the Hukaung Valley to survey the rattans. We completed the survey in early 2005, discovered a new species of rattan, and came to appreciate that rattan was just one of many forest resources that the communities in the Hukaung Valley depend on. I returned to the Hukaung Valley in April 2009 to do a preliminary resource survey in three villages, and again in December of that year to set up a one-hundred-hectare intensive management area in the forests outside the village of Shinlonga. Shortly after we finished this work, the political situation in Kachin State started to deteriorate and the community forestry work had to be put on hold.

After a four-year hiatus, collaborative community forestry work with the WCS was reinitiated in 2013 in a small village in the Naga Self-Administered Region. This village presented an interesting case study because of its long history of using and managing forest-lands that were outside the jurisdiction of the Forest Department. I expanded the community-based natural resource management work in Myanmar during several visits in 2014, collaborating in household surveys, making preliminary inventories in the village of Nam Sabi in Sagaing Region, and helping to set up a one-hundred-hectare management area in the forests outside the village. The Nam Sabi village management area (VMA) is located in the buffer zone of the Htamanthi Wildlife Sanctuary. Because of its location along the periphery of a wildlife sanctuary, the Nam Sabi VMA represents a valuable example of involving local communities in the conservation and sustainable use of protected areas. The policy implications here are huge. I have continued to collaborate with my friends at the Wildlife Conservation Society for over ten years now, and we have accomplished some marvelous things in Myanmar. And it all started with one phone call.

There were twelve of us on the initial rattan survey of the Hukaung Valley in January 2005, including three local botanists, field assistants, and cooking staff. All our bags, tents, cooking equipment, food, and plant presses were loaded into two monster trucks with large knobby tires and an exhaust pipe routed up along the front windshield so that the trucks could cross rivers and sustain water up to the doors without the motor stopping. Six of us crowded in the back of each truck, and alternated standing and sitting (and sleeping) as the trucks slowly worked their way up the Ledo Road to the little town of Namyun on the border with India.[1] We crossed several deep rivers along the way without any difficulties thanks to the exhaust pipe.

It started pouring rain soon after we arrived at Namyun. The crew quickly unpacked the trucks and moved our gear into the school building; we set up our tents inside and strung a clothesline to start drying wet clothes. The trucks headed back down the Ledo Road in the rain, and the cooks served us a hot meal by candlelight on the teacher's desk. Conversation was difficult with the rain banging on the tin roof of the school, but we discussed our schedule for the next six weeks, explained to the local botanists the general plant collecting and inventory methodologies we would use, and speculated as to when the elephants, which were to carry our equipment, might arrive.

The mahouts arrived with their elephants the next day. All the gear that had been in two trucks was transferred to the backs of two elephants and carefully lashed into place. We would walk alongside the elephants for five or six hours until we came to a good place to make camp, then set up our tents and a temporary kitchen.[2] We would spend several days at each camp while we explored the local forests, collecting samples of rattans with flowers or fruits, and running inventory transects to document the density and size distribution (number of seedlings, saplings, juveniles, and adults) of local rattan populations. These data can show whether a species is regenerating itself in the forest and periodically establishing new seedlings and saplings or whether the population contains only large adult canes, no seedling or saplings, and will eventually disappear from the forest.

Immediately after returning from the field each day, the botanists would trim their plant collections to fit on herbarium sheets, fold them in newspapers, and stuff them into thick plastic garbage bags. Before tying up the bag, they would pour in several cups of a 70-percent alcohol solution to preserve their specimens. We usually had dinner—rice, vegetables, peanuts, spicy hot tofu—sitting outside on a log or little plastic stools, but if it was raining, we would pile into the kitchen, where the smoke would be so thick that it burned our eyes. People didn't linger over dinner on rainy nights. The research team would retire to their tents to read or write up their journals, while the kitchen crew washed dishes and the field assistants sharpened their bush knives. The mahouts would untether the elephants and let them forage through the forest during the night. Lying in my sleeping bag, I could hear the bell tied around the elephant's neck gently ringing as the animal bent down to grab a leaf, the sound growing fainter and fainter as the elephants moved farther into the forest. They would always be back at camp the next morning.

We continued this rhythm—walking, making basecamps, collecting plants, and running inventory transects—for four weeks. When we finally walked out of the Hukaung Valley into the town of Tanai, where we planned to spend several days drying plant specimens at the headquarters of the Tiger Reserve, we had eight garbage bags stuffed with plant specimens reeking of alcohol. We had collected over two hundred specimens, including fifteen species of rattan; eight of the rattan species had never been recorded in Myanmar previously, and one of the rattans was a new species to science and now bears the name of the Hukaung Valley: *Calamus hukaungensis* A.J. Hend. (Arecaceae). We had run ten transects at different elevations along the Ledo Road and collected quantitative data on the structure of wild populations of twelve rattan species. Some of the commercial species in the reserve occurred at densities of over a hundred harvestable canes per hectare.

In spite of intense commercial harvesting, the rattan populations along the Ledo Road were thriving. Because of the low price offered for the cane, collectors rarely venture more than a kilometer from the road to look for rattan, and owing to the noise, the dust, and the presence of people, tigers rarely come within a kilometer of the road. We concluded that tiger conservation and rattan collection could coexist, and that promoting the management and sustainable exploitation of the latter by local communities would greatly enhance existing efforts to conserve the former. Developing sustainable harvest levels for commercial rattans in the Hukaung Valley, for instance, would ensure that collectors could find sufficient quantities of merchantable cane to exploit near the Ledo Road; they would thus have little incentive to go farther into the forest. Fixed harvest quotas would also help dampen the boom-and-bust dynamics that characterize the markets of many wild harvested resources. Collectors could count on a reliable income from harvesting rattan each year, and their encounters with tigers deep in the forest would become less frequent.

Before heading up the Ledo Road, the kitchen crew had bought plastic cups for everyone on the expedition. They were all different, and mine was white with flowers and a tight-fitting lid for keeping tea hot. I wrote my name, the date, and the location on the bottom of the cup. After driving for several hours, we stopped to stretch our legs and eat lunch at mile 18 camp. We ate curry and drank tea, and talked to the locals about rattan. When we arrived in Namyun, I could not find my cup; I had left it at my host's house at mile 18. One of the kitchen crew graciously gave me his cup—hot pink with no lid. I was grateful, but disappointed that I had lost my cup on the first day.

The next morning, one of the cooks approached me with a big smile and handed me a blue plastic bag with my cup inside. The woman of the house had noticed that I had left my cup on the table, and she knew that we were headed to Namyun. She gave it to the first motorcyclist who stopped on the way to Namyun with instructions to give it to the group with the American. I have encountered similar instances of careful attention to detail and kindness by the Myanmar people throughout my fieldwork. My overall enthusiasm for community-based approaches to resource management in this country is largely the result of a long series of wonderful interactions with local villagers.

We put together a copiously illustrated report for the Myanmar Forest Department soon after returning from our Hukaung Valley trip and started planning a follow-up project in the tiger reserve. We knew that subsistence communities living inside the wildlife sanctuary depended on the forest for timber, bamboo, and rattan to build their houses, for palm thatch, and for medicinal plants, but we did not know which species they preferred, or how much they used, or what impact the exploitation of these products was having on the forest. Our project proposal was eventually approved, and in May 2009 we conducted preliminary surveys in three villages in the central part of the reserve, Shinlonga, Lajarbon, and Takhet, to get a better idea of what could be achieved with a collaborative community forestry initiative.

We set up a base camp in Shinlonga, a small Kachin community, and conducted household interviews and ran inventory transects at each village. The interviews were usually done in the evenings after the workers had come back from the fields. We would sit around the glow of the cooking fire in the center of the house, and with our flashlight illuminate different sections of the structure—the floor, the walls, the roof joists, the roof—and ask what each was made of, where the material had been harvested, and how long it lasted. It took about half an hour to do the interview and, like people I have encountered all over the world, the inhabitants seemed to enjoy talking about their house, especially if they had built it themselves.

All the houses had thatch roofs, with the cooking fire centrally located inside the house. The underside of the thatch above the fire was always black from the smoke, but the villagers explained that the smoke helped waterproof the roof so it would last longer. Every house had a long bamboo pole leaning against the roof outside with a large square of plaited bamboo tied to the end like a huge flyswatter. This, I was told, was used to put out roof fires, though many houses had a bucket of water handy for the same purpose. Kachin houses require a lot of maintenance. The palm thatch needs to be replaced every three to five years, and the plaited bamboo walls last only about ten years.

Our inventory transects revealed that most of the important timber trees were present in the forest solely as pole-sized individuals, and many of the commercial rattan species had been completely harvested out. But we knew from the household surveys that villagers were not harvesting either of these resources in large quantities. When asked about this, they reported that commercial timber and rattan concessions had been granted in the forest—inside the wildlife sanctuary—and trucks from Tanai had come several times to harvest sawlogs and collect rattan. Many of the Kachin communities had been moved out of the forest to sites along the road so that the Myanmar Armed Forces could "keep an eye on them." They had been given use rights to the forests surrounding their villages as an incentive to relocate, but several years later, outside concessions had been granted for the same piece of forest. Community forestry would never work as long as it was not clear who had control over the resources.

Good palm thatch was also becoming increasingly hard to find. The preferred roofing material in the Hukaung Valley is produced by the *tawhtan* palm (*Livistonia jenkinsiana*).[3] The palmate leaves of this species can last up to five years, and the amount of thatch that can be collected from one individual far surpasses that of any other palm. Traditionally, the Kachin would plant tawhtan palms along the perimeter of their rice fields and harvest the leaves as they needed them. Few of the communities had done this since moving to the Ledo Road, and most householders currently harvested thatch from wild palms, whose leaves are less durable and need to be replaced more frequently. Nobody wanted to commit to planting around their fields for fear of being forced to move before they could harvest it.

The socioeconomic conditions in Shinlonga, as well as the ecological conditions in the forest surrounding it, seemed ideal for a program of community-based natural resource management. The community was enthusiastic about the prospect and in late 2009 selected a hundred-hectare tract of forest located about two hours north of the village near a small river for their management area. As with the project I had done previously in Brazil, we would lay out a baseline and then conduct a 10-percent inventory of the entire area using parallel transects. We walked to the management area the first day, but we were so tired by the time we got back to the village that we decided to make a basecamp the next morning. While the cooks saw to that, the field crew and the science staff—U Zaw Lin, U Saw Htun, Rob Tizard, and I—walked directly to the management area and worked on the baseline, clearing the line, tying flags every twenty meters, setting a painted transect post every hundred meters, and obsessively making sure that the line was lying straight. When we arrived, exhausted, at the basecamp site later that afternoon, we found a large, thatch-roofed hut for the field crews, a kitchen with fire pit, and three tents set up next to a beautiful bend in the river. One of them had my pack and sleeping bag inside.

Naw Aung is a Kachin who helped with the forestry work during both my visits to Shinlonga. A major asset in the field, he could identify all the trees, was the first to cut a plot stake or grab a transect rope or start clearing a line, and always had a big smile on his face. He wore an old army shirt, a *longyi* (a piece of cloth sewn into a cylinder worn around the waist like a sarong), flip-flops, and a large conical hat, and would light up a cheroot whenever he took a break. The handle of Naw Aung's bush knife cracked once, probably from overuse, while we were laying out the baseline of the management area. He repaired it by the fire that evening using a local rattan,[4] and I was not the only person who was mesmerized while watching him do it. He first split the rattan cane and shaved it down to the appropriate thickness. Then he wove a ring of rattan, which he slipped over the cracked handle and pounded into place to close the crack using another piece of rattan. He finished the job by weaving a beautiful rattan sleeve around the handle, and then oiled it using grease from the kitchen. The bush knife was back in use clearing line the next day.

The crews completed all twenty transects in the management area and collected a phenomenal amount of data. We now knew how much timber, bamboo, rattan, palm thatch, and medicinal plants grew in the reserve, and we were able to make contour maps showing the spatial distribution of different resources in the area. These could later be used to establish harvest zones. I put in an order for several thousand stainless-steel springs and a hundred rolls of metal strapping for the growth studies of the timber trees. We were close to establishing the first verifiably sustainable, data-driven participatory community-based forest management project in Myanmar. But something else was happening at this same time—the Kachin Independence Army, the military wing of the Kachin Independence Organization (KIO), a political group of ethnic Kachin who have been fighting for their independence—and more recently their autonomy—since the early 1960s, renewed fighting with the Myanmar Armed Forces, civil war broke out, and all research teams and nonresidents were pulled out of Kachin State.

Four years went by, but the situation in Kachin State remained dangerous and complicated, so we decided to investigate the possibility of doing community forestry work in another location. Tikon, a small village of mixed Naga-Chin ethnicity in the Naga Self-Administered Zone near the border with India was recommended because of its unusual tenure context—the community had lived for a hundred years in a huge tract of forest that was not directly controlled by the Forestry Department.[5] It sounded promising so we flew to Mandalay, then took another flight to Homalin, then got on a boat going up the Chindwin River. Six hours later, we got off at Htamanthi and climbed into a truck to drive to mile 25 camp, where there were motorcycles waiting to carry our bags—and maybe us—the final ten miles through the mountain forest to the village. But it started raining and we had to walk because the motorcycles were having trouble getting up the hills. We arrived at the village ahead of our bags—and our dry clothes—just as the rain stopped. The village, consisting of thirteen houses, a church, and a school, was shrouded in fog, a few rays of brilliant sunlight streaming through the clouds. We were at an elevation of over 1,500 meters, the air was brisk, and the Naga Hills and the Indian border loomed in the background. The motorcycles with our bags finally arrived and we pitched our tents in the wood-frame schoolhouse.

In spite of a hundred years of continual use, the montane forests surrounding the village were extremely well preserved—multistoried, closed canopied, with numerous big trees, understory palms, and tall, looping rattans; our inventory transects were usually accompanied by the raucous singing of Hoolock gibbons (*Hoolock leuconedys*).[6] The traditional territory of the village encompasses over two thousand hectares of forest, and the villagers have managed it well. Hundreds of tall, straight timber trees can be found there, as well as an abundance of good bamboo and rattan canes, and a surprising richness of medicinal plants. With good reason, the villagers worry that someone will take these resources, as well as the forests that contain them, away from them.

We put growth bands on some of the timber trees to see how fast they were growing, and held a village meeting to explain what we had found out in our household surveys and forest inventories. The entire village came to the meeting. We discussed the possibility of developing a community forestry project with the village, and everyone seemed enthusiastic about the idea. We made it clear, however, that this project would involve the intensive management of a *small* tract of forest, with the objective of developing a management plan and applying on behalf of the village for a community forestry certificate, a permit issued by the Myanmar Forest Department that provides villagers with usufruct rights to specific forest resources, or plantings, or, in selected cases, intact forests. Such a permit would guarantee the villagers the rights to that piece of forest alone. We explained that given the current economic realities in Myanmar, the chances of the community maintaining control over two thousand hectares of resource-rich forest were pretty slim.

My first morning in the Naga Hills, as I sat on the front porch of the schoolhouse writing in my journal, I saw two policemen with automatic weapons and walkie-talkies walking into the village. I scurried inside, and listened to the conversation out in the schoolyard; then the village head took the two men over to his house for breakfast. I found out later that the two policemen had been sent over from the district capital to make sure that the two foreigners (Rob Tizard and I) were "all right."

The policemen, and their guns, went with us to the field to observe what we were doing. They also participated in some of the tree banding exercises and even put their guns down once to help us make growth bands. A light drizzle was falling the day we walked out of the village back to mile 25 camp, and the red clay on the road was treacherous. I had a walking stick, but I was proceeding with extreme caution as I worked my way down the first slope. One of the policemen came up from behind me and gently took my hand. When we got to the bottom of each hill, and I was no longer in danger of slipping, he would release my hand and drop back next to his partner. He did this every time we had to walk down a hill, for ten miles, all the way back. I never fell. We made brief eye contact after we arrived at mile 25 camp and he gave me a little smile.

Following each of my trips to Myanmar, I would write a detailed report of my activities for the Forest Department. Composing these reports was a useful exercise that enabled me to analyze data and plan the next steps in our community forestry work. I would always include an array of photographs and make the reports as appealing as possible—but I never knew whether anybody actually read them or whether they went straight into a file.

In May 2014, I had the opportunity to go to the capital city, Nay Pyi Taw, and meet the director general of the Forest Department, Dr. Nyi Nyi Kyaw. My visit was mainly a courtesy call; I wanted to thank him for supporting our work over the previous ten years and tell him about what we were currently doing. We had a lovely meeting, and talked about rattan, the importance of botanical exploration, and community forestry. Dr. Nyi Nyi Kyaw said that he was convinced that engaging with rural communities was the quickest, most effective way to promote the conservation and sustainable use of Myanmar's forests. But such a procedure was unlike the way the Forest Department had operated in the past, and they were still trying to determine how best to move forward with community-based natural resource management. He then casually dropped into the conversation that *he had noticed from reading my reports* that I had been developing some interesting community forestry prototypes in northern Myanmar. He asked me to keep him informed on how this work was going.

After ten years of fieldwork in northern Myanmar, my WCS collaborators and I had come close to developing a community forestry program for a village located inside a protected area, and we had done preliminary resource supply-and-demand analyses in a community located on forestlands outside of the jurisdiction of the Forest Department. The final land-tenure issue left to investigate was the situation of communities located in the buffer zone of a protected area. These communities have the potential to make a large contribution to conservation. If a program of sustainable forest use could be implemented in the forests surrounding a protected area, and if the community were able to assume control over the forest and stick to the management prescriptions, harvesting the resources they needed for subsistence and perhaps selling a few others to provide a sustainable source of income, the villagers would have little incentive to encroach on the core area of the reserves and poach wildlife. A viable program of community-based natural resource management would, in the words of the director general, "help to lock down the buffer zone and create a sense of stewardship at the village level."

We found what looked like an appropriate community, Nam Sabi, on the banks of the Chindwin River in Sagaing Region. The traditional territory of this small Shan village extended into the buffer zone of the Htamanthi Wildlife Sanctuary, a two-thousand-square-kilometer wildlife reserve located between the Chindwin and Uyu Rivers. The sanctuary contains tigers, elephants, leopards, and bears. We started working with this village in May 2014, and finished all the preliminary household surveys and forest inventories. We discussed with the community the prospect of setting up a village management area and applying for a Community Forestry Certificate. We found a lot of support for the idea, and on my next trip in early November 2014 we built a basecamp out in the forest and started laying out a hundred-hectare management area. It took two days to set all the transect stakes along the baseline, and we were able to finish two transects before I left. The remaining eighteen transects were completed by village crews under the direction of U Myint Thein, an extremely capable ranger from the Htamanthi Wildlife Sanctuary with whom I had worked on two previous trips. Not only did the field crews finish all the transects, but they finished before the end of November in time to harvest their rice fields. They also put growth bands around eighty-two timber trees to start growth studies.

The community forestry work in Sagaing Region has surpassed the work that was interrupted five years ago in the Hukaung Valley. The site and context are different, but the underlying concept is the same—balance the local supply and demand for forest resources at the community level through the sustainable harvest and participatory management of a small tract of forest. A lot of enthusiasm exists at the village level to make this work, giving me a strong feeling that a new story about people and plants and tropical forests is being born here.

When we finished running the first transect in the Nam Sabi management area, it was about 2:00 p.m. and past time for lunch. We were close to a large stream that ran through the area, and the smooth, sandy beaches seemed perfect for a picnic. We hiked over to the river and had a delightful lunch, relaxing, talking about the transect we had just finished, and marveling at the beauty of our lunch spot. The next day, when we finished transect two, we decided to walk back to the river where we had had lunch the day before. Clearly outlined in the damp sand, fresh from the night before, we saw tiger tracks. I am sure the tiger had smelled us. I have never wanted to meet a tiger in the forest, but I feel deeply honored to have shared a resting spot with one.

Epilogue

Stories can mean different things to different people. A forester, an ecologist, or a farmer might not interpret a story in the same way, and each of them might read it differently from an anthropologist, a lawyer, or a poet. Regardless of the author's intention, once a story drops into the life of its readers, each with an individual history, outlook, and set of core beliefs, it will be subtly changed. And each reader will accept or reject its conclusions, in whole or in part, on the basis of his or her current worldview. Stories inform our lives and constitute our world, but speculation about what they mean can be complicated.

Monographs often end with a section in which the author pulls everything together, drawing conclusions from the preceding chapters and furnishing lessons for the reader. I would like to offer here something less concrete; rather, I shall sketch out some of the general patterns I have noted about people, plants, and tropical forests from the point of view of a person who was a participant in all these stories.

Epilogue

In spite of the drastic political, social, and environmental changes that have characterized village life in the tropics over the past decades, it is noteworthy that local communities retain a deep knowledge about the properties and uses of their tropical forest resources. They can identify important species, and they know what these are used for, where they grow, and how much of them can be sustainably harvested.

I have been continually impressed by the management prowess of indigenous foresters. These skilled practitioners, mostly full-time farmers, have developed a variety of extremely sophisticated, parsimonious, and cost-effective systems for managing tropical forest resources. They have been doing this for a very long time, and they are much better at it than industrialized Westerners are. Forest-dwelling communities have been enriching forests with useful species and subtly controlling the successional development of forest vegetation throughout the tropics. As can be guessed, the title of this book was chosen to be deliberately provocative. There is, truth be known, no "wild" in the tropics, because local communities have been doing forest management for thousands of years. If Westerners want to learn about the conservation and sustainable use of tropical forests, these are the people we need to talk to first.

A common thread running throughout many of the stories is that villagers, even those with limited formal education, can be taught to collect the baseline data required by local forest departments for the permits needed to "officially" manage their forests. The work involves quantifying the more qualitative and subjective forest characteristics that have traditionally guided their management operations, such as putting numbers on resource density and yield, making graphs, and packaging the data in the appropriate format. I have noted that everyone in a village appreciates the benefits of counting and measuring, as opposed to visual estimates and educated guesses, in making decisions about which resources to harvest.

My fieldwork has shown me that the community management of tropical forests is what villagers do, routinely, all over the world, and that it is as much a part of village life as collecting firewood, planting a field of corn or rice, or harvesting palm thatch to roof a house. All these activities, in fact, are probably components of the local management system. But outsiders rarely see these linkages, or associate swidden (slash-and-burn) farming with forestry, or appreciate how periodic slash weeding can change the composition and structure of a tropical forest. Community forestry is largely invisible on land-use maps and tables of annual statistics produced by central governments. As a result, beyond saying that "probably a great deal" of tropical forest is managed and conserved by local communities, we know very little about the areal extent of this work. I would venture that the expanse is much bigger than we might imagine—but that it is getting smaller each year.

Finally, it strikes me that the stories offered here provide a useful counterpoint to the generally bleak and dispiriting reports coming out of the tropics, and suggest that much of what Westerners generally assume about indigenous people and their management of tropical forests is incorrect. At this juncture of global climate change, accelerating species extinctions, and burgeoning human populations, we must together find better ways to manage tropical forests than to degrade them or cut them down. Based on over thirty years of fieldwork, I am convinced that the best suggestions will come from the people who live in these forests.

Notes

INTRODUCTION
The Challenge of Sustainable Forest Use

1. Tea, *Camellia sinensis* (L.) Kuntze, for example, typically grown in plantations, also occurs wild as large, scattered forest trees in northern Myanmar.
2. Rattans are a group of approximately six hundred species and thirteen genera of climbing, spiny palms native to the tropical forests of Asia, Africa, and Australasia; the stems are used for making furniture and handicrafts. Rattan collecting is an important part of the livelihoods of many rural communities, as I discuss in Chapter 14.
3. For other approaches see M. N. Alexiades, C. M. Peters, S. A. Laird, C. Lopez-Binnquist, and P. Negreros-Castillo, "The Missing Skill Set in Community Management of Tropical Forests," *Conservation Biology* 27 (2013): 635–637; C. M. Peters, M. N. Alexiades, and S. A. Laird, "Indigenous Communities: Train Local Experts to Help Conserve Forests," *Nature* 481 (2012): 443; C. M. Peters, *The Ecology and Management of Non-Timber Tropical Forest Resources* (Washington, D.C.: World Bank, 1996); and C. M. Peters, *Sustainable Harvest of Non-Timber Plant Resources in Tropical Moist Forests: An Ecological Primer* (Washington, D.C.: Biodiversity Support Program, 1994).

ONE
The Ramón Tree and the Maya

1. *Chicleros* are people who climb *chicle*, *Manilkara zapota* (L.) P. Royen, trees and tap them for latex, which was traditionally used for making chewing gum.
2. Full name: *Brosimum alicastrum* Sw. (Moraceae). Throughout, I shall give only the binomial in the text and supply the full scientific name in the notes.
3. See C. M. Peters and E. Pardo-Tejeda, "*Brosimum alicastrum:* Uses and Potential in Mexico," *Economic Botany* 36 (1983): 166–175.
4. El Tajin was named a UNESCO World Heritage Site in 1992.
5. See the image accompanying Chapter 11 of a forester installing a growth band.
6. This type of breeding system is termed *gynodioecious.* The population has female and hermaphroditic trees; the male condition is derived from hermaphroditic trees. See C. M. Peters, "Plant Demography and the Management of Tropical Forest Resources: A Case Study of *Brosimum alicastrum* in Mexico," in *Rain Forest Regeneration and Management*, ed. A. Gomez-Pompa, T. Whitmore, and M. Hadley (Carnforth: Parthenon, 1991), 265–272.
7. I discuss this in C. M. Peters, "Observations on Maya Subsistence and the Ecology of a Tropical Tree," *American Antiquity* 48 (1983): 610–615; and C. M. Peters, "Pre-Columbian Silviculture and Indigenous Management of Neotropical Forests," in *Imperfect Balance: Landscape Transformations in the Pre-Columbian Americas*, ed. D. Lentz (New York: Columbia University Press, 2000), 203–224.
8. Tikal is located in the Petén Basin of northern Guatemala. Population estimates for Tikal range from ten thousand to ninety thousand inhabitants; the ruins have been completely mapped, and there appear to be over three thousand structures.
9. Tree species with male and female flowers on the same individual are called *monoecious.*
10. The "ecological niche" of a plant species can be defined in more than one way, but it is frequently described as the combined *abiotic* (that is, through nutrients, water, and light) and *biotic* (by means of predators, pollinators, and pathogens) conditions that allow a particular species to grow and reproduce. That tropical forests contain so many different types of plants would suggest that these habitats offer an abundance of niches for species to exploit.

TWO
Mexican Bark Paper

1. At its peak, Tenochtitlán, which was conquered by Hernán Cortés in 1521, was the largest city in the Americas.

2. See C. M. Peters, J. Rosenthal, and T. Urbina, "Otomi Bark Paper in Mexico: Commercialization of a Pre-Hispanic Technology," *Economic Botany* 41 (1987): 423–432.
3. Full name: *Trema micrantha* (L.) Blume (Ulmaceae).
4. I used a replicated, factorial (2 x 4) experiment to examine the effect of type of covering and intensity of bark removal on bark regeneration. The experiment included two types of covering, aluminum foil and banana leaves, and four intensities of bark removal—20 percent, 40 percent, 60 percent, and 80 percent—and was replicated three times.
5. A cork borer was used to collect samples of the bark and the trunk of the harvest trees. These samples were then thinly sliced with a microtome, mounted on slides, and examined under a microscope.

THREE
Camu-camu

1. *Myrciaria dubia* (Kunth) McVaugh (Myrtaceae) is a shrub or small tree native to western Amazonia, and it is a common component of riparian vegetation along the Nanay, Napo, Ucayali, Marañon, and Tigre Rivers in Peruvian Amazonia.
2. At the time this research was done, the city of Iquitos was located on the left bank of the Amazon River and had a bustling port where the riverboats would dock and unload their freight. There is still a bustling port, but it is no longer located on the left bank of the Amazon River. The rivers in Amazonia move around a lot.
3. Chaucheros carry cargo to and from the riverboats. Using only a tumpline, they carry motorcycles, fifty-five-gallon drums, gunnysacks full of rice, and baskets of camu-camu fruit up and down the riverbank at the port in Iquitos.
4. Oxbow lakes are U-shaped bodies of water that are formed when the wide meander of a river is cut off. River meanders carve out soil on the outside of the curve and deposit soil on the inside of the curve. In time, the river will cut completely through the curve, forming a straight channel and leaving the curve behind as an isolated lake. Oxbow lakes are called *cochas* in Peruvian Amazonia.
5. The fieldwork described in this chapter was conducted at Sahua cocha.
6. The fruits of camu-camu are an important food source for many species of fish, including the commercially valuable *gamitana* (*Colossoma macropomum* G. Cuvier).
7. We published them in C. M. Peters and E. J. Hammond, "Fruits from the Flooded Forests of Peruvian Amazonia: Yield Estimates for Natural Populations of Three Promising Species," *Advances in Economic Botany* 8 (1990): 159–176.

8. See M. P. Martin, C. M. Peters, and M. S. Ashton, "Revisiting Camu-camu (*Myrciaria dubia*): Twenty-Seven Years of Fruit Collection and Flooding at an Oxbow Lake in Peruvian Amazonia," *Economic Botany* 68 (2014): 169–176.
9. Full name: *Eugenia inundata* D.C. (Myrtaceae).

FOUR
Fruits from the Amazon Floodplain

1. Full name: *Grias peruviana* Miers (Lecythidaceae).
2. Color, origin, and sediment load are used to classify the different types of rivers in Amazonia. Whitewater rivers, like the Amazon and the Ucayali, are full of suspended sediments that are brought down from the surrounding mountains. Clearwater rivers, like the Xingu and Tapajos, have a bluish color and carry few sediments because of their origins in ancient rock formations like the Guyana Shield that no longer erode. Blackwater rivers, like the Rio Negro, are extremely low in sediments, dissolved nutrients, and bacteria—a bit like distilled water—and the pH of the water is acidic. These rivers have their origins in lowland tropical forests.
3. I used matrix models, computer simulations, and eigen values often when I first started working in tropical forests. I eventually reached a point of diminishing returns, scarce computer facilities, and questionable utility.
4. Full name: *Mauritia flexuosa* L.f. (Arecaceae)
5. See C. M. Peters, "The Ecology and Economics of Oligarchic Forests," *Advances in Economic Botany* 9 (1992): 15–22.
6. See C. Padoch, "Aguaje (*Mauritia flexuosa* L.f.) in the Economy of Iquitos, Peru," *Advances in Economic Botany* 6 (1988): 214–224.
7. This breeding system, which is exhibited by a relatively small percentage of tropical tree species, is called *dioecy*.
8. Full name: *Euterpe oleracea* Mart. (Areceaceae).
9. Full name: *Euterpe precatoria* Mart. (Arecaceae).
10. Recall that meristems are undifferentiated tissues in plants that divide and give rise to various organs. The *apical meristem* gives rise to leaves, flowers, and new shoot tissue, while *lateral meristems*, like the vascular cambium described in Chapter 2, give rise to the xylem and phloem that allow a plant to grow in diameter. Palms possess only a single apical meristem and do not grow in diameter.
11. An oligarchic forest dominated by a single species is sometimes referred to as a monospecific forest.
12. Full name: *Spondias mombin* L. (Anacardiaceae).
13. Materos are specifically locals who can identify the trees in the forest. Some Amazonian materos can also identify the trees by their scientific names.

14. Full name: *Genipa americana* L. (Rubiaceae).
15. C. M. Peters, A. H. Gentry, and R. O. Mendelsohn, "Valuation of an Amazonian Rainforest," *Nature* 339 (1989): 655–656.
16. See *New York Times*, July 4, 1989, and *Washington Post*, June 28, 1989; the front page of the *Post*, below the fold, has a photo of a local harvester in his canoe with a load of camu-camu fruit (see Chapter 3).

FIVE
Forest Fruits of Borneo

1. Full name: *Shorea atrinervosa* Sym. (Dipterocarpaceae)
2. My results are published in C. M. Peters, "Illipe Nuts (*Shorea* spp.) in West Kalimantan: Use, Ecology, and Management Potential of an Important Forest Resource," in *Borneo in Transition: People, Forests, Conservation, and Development*, ed. C. Padoch and N. Peluso (New York: Oxford University Press, 1996), 230–243.
3. This number may change; not all the collections from the site have been positively identified even now.
4. The oleo-resin produced by several genera of trees in Southeast Asia is known as *damar*. The dried exudate is a valuable source of varnish and caulking. Indigenous communities in Borneo and the Malay Peninsula harvest it by climbing the trees and cutting small pyramidal holes or boxes at various heights along the trunk. In response to the wounding, the tree starts to exude oleo-resin, which collects at the bottom of the boxes; a bush knife is used to scrape out the material. The flow usually stops after a couple of days, when the cells become clogged with dried resin. To start the resin flowing again, twigs and dried leaves are used to make a small fire in each box. A large buttressed damar tree dotted with dozens of little fires burning along its trunk is an unforgettable sight. This method of tapping plant resins is known as boxing and firing.
5. The best bush knives, I am told, are forged from the leaf springs of a Toyota Land Cruiser.
6. Full names: *A. heterophyllus* Lam.; *A. altilis* (Parkinson) Fosberg; *Artocarpus integer* Merr.; *Mangifera* spp. (Anacardiaceae); *Nephelium* spp. (Sapindaceae); *Parkia speciosa* Hassk. (Fabacaeae); *Durio zibethinus* Rumph. ex Murray (Malvaceae); *Garcinia mangostana* L. (Guttiferae); *Salacca zalacca* (Gaertn.) Voss (Arecaceae); and *Baccaurea* spp. (Euphorbiaceae).
7. If I remember correctly, Christine Padoch said this.
8. A becak has two wheels in front and one in the back. The driver sits in the back and the passengers sit in the front. Two or three adults, and an almost inexhaustible number of children, can cram into the front seat of a becak.

9. Full name: *Eonycteris spelaea* Dobson (Pteropodidae).
10. Full name: *Sonneratia alba* Sm. (Lythraceae).
11. Full name: *Lansium domesticum* Correa (Melicaeae).
12. Botanically, the fleshy covering surrounding the seed is known as an *aril.* The edible parts of a rambutan, mangosteen, and durian fruit are also arils.
13. My description of the agroforestry system developed in Punggur for use in peat soils is based on data collected by Elysa Hammond.
14. The rice is transplanted twice to produce a seedling with a long root system. A two-meter-long dibble stick is used to punch a hole through the peat so that the roots of the rice seedling make contact with the mineral soil.

SIX
Homemade Dayak Forests

1. The official language of Indonesia, a variant of Malay, is Bahasa Indonesia. Since Indonesia is the fourth-most populous country in the world, and most Indonesians speak Bahasa, it is one of the most widely spoken languages in the world. An additional seven hundred or so indigenous languages are spoken throughout the Indonesian archipelago.
2. Full name: *Arenga pinnata* (Wurmb.) Murr. (Arecaceae).
3. The Dipterocarpaceae is a dominant family of tropical trees in Southeast Asia comprising seventeen genera and about five hundred species. Dipterocarps are trees that belong to the family Dipterocarpaceae.
4. Full name: *Eusideroxylon zwageri* Teijsm. & Binn. (Lauraceae).
5. The villages, shown in parentheses, contained different Dayak groups: Selako (Bagak Sahwa), Tara'n (Tae), Iban (Empaik and Tematu), and Kenyah (Ensibau).
6. These exchanges were conducted in Bahasa, or were translated from the local Dayak language by bilingual field assistants.

SEVEN
Sawmills and Sustainability in Papua New Guinea

1. Communities (*ejidos*) also own a large percentage of the tropical forests in Mexico.
2. Portable sawmills usually have a band saw, rather than a circular blade, which moves back and forth while the log stays in one place.
3. Full name: *Metroxylon sagu* Rottb. (Arecaceae).
4. This was during the early days of GPS, and the receivers were big and clunky and did not work very well under a closed canopy.

5. While the inventory work at Kikori was focused on timber species, other community forestry inventories that I have been involved with have also included palms, climbers, and selected species of medicinal plants.
6. Full names: *Xylocarpus granatum* J. Koenig (Meliaceae); *Intsia bijuga* (Colebr.) Kuntze (Fabaceae); *Calophyllum papuanum* Lauterb. (Calophyllaceae).
7. Several of the Kikori inventories involved our wading though water up to our chests while holding a clipboard with the field notes on top of our heads. While the water was somewhat refreshing because of the heat, I was always a little worried about what I couldn't see and what I might be stepping on.

EIGHT
Collaborative Conservation in the Bwindi Impenetrable Forest Reserve

1. Pit sawing involves positioning a log over a pit and using a long, two-handled crosscut saw to cut it into boards. One of the sawyers stands on top of the log, the other is down in the pit.
2. Full name: *Gorilla beringei beringei* Matschie (Hominidae).
3. Full name: *Loxodonta africana* Blumenbach (Elephantidae). This elephant is the largest living terrestrial animal.
4. Full names: *Eleusina indica* (L.) Gaertn. (Poaceae) and *Plantago palmata* Hook. f. (Plantaginaceae).
5. Full name: *Loesneriella apocynoides* (Welw. Ex Oliv.) J. Raynal (Celastraceae).

NINE
A Renewable Supply of Carving Wood

1. Full name: *Bursera glabrifolia* (H.B.K.) Engl. (Burseraceae).
2. See C. M. Peters, S. A. Purata, M. Chibnik, B. J. Brosi, A. M. López, and M. Ambrosio, "The Life and Times of *Bursera glabrifolia* (H.B.K.) Engl. in Mexico: A Parable for Ethnobotany," *Economic Botany* 57 (2003): 431–441.
3. The forestry group was IXETO, the Union of Forestry Communities in Ixtlán-Etla Oaxaca.
4. Specific gravity is calculated by weighing a cubic centimeter of wood. Using this constant, an estimate of weight (grams or kilograms) can be converted to an estimate of volume (cubic centimeters or meters).
5. The stump is marked with a seal so that after the tree is felled, the stump carries the evidence that the tree was cut legally. A stump without a mark indicates that the tree was not supposed to be cut.

TEN
Caboclo Forestry in the Tapajós-Arapiuns Extractive Reserve

1. Full names: *Bothrops* spp. (Viperidae) and *Boa constrictor* L. (Boidae).
2. Full name: *Chelonoidis denticulate* L. (Testudinidae)
3. See D. G. McGrath, C. M. Peters, and A. Motes Bentes, "Community Forestry for Small-Scale Furniture Production in the Brazilian Amazon," in *Working Forests in the Neotropics*, ed. D. Zarin, J. Alavalapatti, F. Putz, and M. Schmink (New York: Columbia University Press, 2004), 200–220.

ELEVEN
Measuring Tree Growth with Maya Foresters

1. See C. M. Peters, "Pre-Columbian Silviculture and Indigenous Management of Neotropical Forests," in *Imperfect Balance: Landscape Transformations in the Pre-Columbian Americas*, ed. D. Lentz (New York: Columbia University Press, 2000), 203–224.
2. Full name: *Swietenia macrophylla* King (Meliaceae).
3. Full names: *Manilkara sapota* (L.) P. Royen (Sapotaceae) and *Alouatta pigra* Lawrence (Atelidae).

TWELVE
Managing Agave, Distilling Mescal

1. The reproductive strategy of producing flowers and fruits one time and then dying is known as monocarpy, or hapaxanthy (for palms), or semelparity.
2. Tropical dry forests occur in regions with a pronounced dry season and contain a large percentage of deciduous species. The tropical dry forests of Mexico are some of the most species-rich in the world.
3. Tequila is made from *Agave tequilana* F.A.C. Weber (Asparagaceae).
4. *Agave cupreata* Trel. & Berger (Asparagaceae).
5. The scientific name for calabash is *Crescentia cujete* L. (Bignonianceae).
6. The National Crusade Against Hunger is a program sponsored by the Mexican government to reduce hunger and alleviate poverty in rural areas. Nestlé and PepsiCo are the major multinational supporters of the initiative.

THIRTEEN

Landscape Dynamics in Southwestern China

1. Numerous Miao rebellions occurred during the Ming (1368–1644) and Qing (1644–1912) dynasties.
2. Full name: *Cunninghamia lanceolata* (Lamb.) Hook (Cupressaceae).
3. See Z. Lu, C. Peters, M. Ashton, J. Feng, and D. Xue, "The Effect of Forest Tenure on Forest Composition in a Miao Area of Guizhou, China," *Mountain Research and Development* 36 (2016): 193–202.
4. Full name: *Pinus massoniana* Lamb. (Pinaceae).

FOURTEEN

The World of Rattan

1. Some of my results are published in C. M. Peters and W. Giesen, "Balancing Supply and Demand: A Case Study of Rattan," *Borneo Research Bulletin* 31 (2001): 138–149; C. M. Peters, A. Henderson, U Myint Maung, U Saw Lwin, U Tin Maung Ohn, U Kyaw Lwin, and U Tun Shaung, "The Rattan Trade of Northern Myanmar: Species, Supplies, and Sustainability," *Economic Botany* 61 (2007): 3–13; and A. Henderson, C. M. Peters, U Myint Maung, U Saw Lwin, U Tin Maung Ohn, U Kyaw Lwin, and U Tun Shaung, "Palms of the Ledo Road, Myanmar," *Palms* 49 (2005): 115–121.
2. A reasonable estimate would be a growth estimate with a standard error of less than 10 percent of the mean value. The sample size needed to achieve this level of precision is determined by a statistical procedure that uses the properties of the variance and mean of the larger data set to estimate the statistical efficacy of progressively smaller sample sizes.
3. During his term as director of the Song Thanh Nature Reserve, Tran Van Thu was pursuing a masters degree at the College of Agriculture and Forestry at Hue University. Because of his close involvement with the rattan work at Song Thanh, he decided to develop a plan for the management and conservation of rattans in the reserve as his thesis project. We awarded him a small student fellowship, and he successfully defended his thesis in early 2012. I believe this was the first formal management plan for rattan—based on voucher specimens, inventory data, and growth studies—ever developed for a protected area in Vietnam. His thesis was the first step toward turning theory into practice.
4. The results were published in C. M. Peters, B. Thammavong, B. Mekaloun, N. Phearoom, O. Ratanak, and T. Ledecq, "Growth of Wild Rattans in Cambodia and Laos: Implications for Management," *Forest Ecology and Management* 306 (2013): 23–30.

5. The new species included two from Cambodia, *Calamus mellitus* Henderson & Khou Eang Hourt, sp. nov. and *Calamus kampucheansis* Henderson & Khou Eang Hourt, sp. nov., and one from Laos and Vietnam, *Korthalsia minor* Henderson & N.Q. Dung, sp. nov.
6. C. M. Peters and A. Henderson, *Systematics, Ecology, and Management of Rattans in Cambodia, Laos, and Vietnam: The Biological Bases of Sustainable Use* (Hanoi: Worldwide Fund for Nature, IKEA, and the New York Botanical Garden, 2014).
7. The forests received both Chain-of-Custody and Forest Certification on over 1,100 hectares of community forest from the Forest Stewardship Council.

FIFTEEN
Community Forestry in Myanmar

1. Under the direction of U.S. General Joseph Stillwell, the Western Allies built the Ledo Road during World War II as a primary supply route to China. The road traverses the entire Hukaung Valley; it is now passable by truck only during the dry season.
2. We made base camps at mile 7, mile 15, and mile 21 along the Ledo Road.
3. Full name: *Livistonia jenkinsiana* Griff (Arecaceae).
4. *Calamus nambariensis* Becc. (Arecaceae).
5. According to the Myanmar Vacant, Fallow, and Virgin Lands Management Law of 2012, the forests surrounding this village are classified as virgin land.
6. Full name: *Hoolock leuconedys* Groves (Hylobatidae).

Acknowledgments

It would be impossible for me to adequately express the deep gratitude that I have for all the people in all the countries who have facilitated my work and provided me with friendship, advice, and shelter. The list is endless, and while the faces and smiles of all are still clearly with me, the names of some have escaped me. To all of you I do not name here, please know that I sincerely appreciate all of your kindness.

For fieldwork in Mexico, I would like to acknowledge the help of Arturo Gómez-Pompa, Enrique Pardo-Tejeda, Lauro López Mata, Rodolfo Dirzo, Daniel Piñero, Silvia Purata, Patricia Gerez Fernández, Guillermo Castilleja, Jorge López-Portillo, José Antonio Sierra-Huelsz, Teodile Urbina, Guadalupe Williams Linera, Gonzalo Castillo-Campos, Andrew Vovides, Mariana Hernández-Apolinar, Myrna Ambrosio, Tarin Toledo, Caterina Illsley, Berry Brosi, Hugo Galletti, Alfonso Arqüelles, Pilar Morales, Jorge Soberón, and Margarita Soto. My work in Peruvian Amazonia was assisted by Umberto Pacaya, José Lopez-Parodi, Armando Vázquez, Miguel Pinedo-Vasquez, Noe Ferry, Rodolfo Vázquez, and Al Gentry. Syamsuni Arman, Wim Giesen, Carol Colfer, Kuswata Kartawinata, Ernst and Hedda Kuester, Pak Afong, Pak Po'on, Ibu and Pak Wim Schouten, Cam Webb, Jim Jarvie, and Li Siang provided invaluable assistance in West Kalimantan. Hank Cauley, Kevin O'Regan, John Barry, and Dembai Auragi played key roles in the Papua New Guinea project. Tony Cunningham offered valuable insights into the impenetrable forests of Uganda. Toby McGrath, Antônio José Mota Bentes, and Dan Nepstad took care of all the logistics in Brazil; Xue Dayuan, Yang Chenhua, Yin Jin, ZhiYao Lu, and Yu Yong Fu provided support in China; and Thibault Ledecq, Bansa Thammavong, Neak Phearoom, Bounchanh Mekaloun, Ou Ratanak, Khou Eang Hourt, Le Viet Tam, Ninh Khac Ban, Biu Van Thanh, and Jack Regalado were key collaborators in the rattan project in Cambodia, Laos, and Vietnam. For the work in Myanmar, I am indebted to Than Myint, Saw Htun, Alan Rabinowitz, Josh Ginsburg, Colin Poole, Rob Tizard, Myint Maung, Saw Lwin, Tin Maung Ohn, Kyaw Lwin, Tun Shaung, Zaw Lin, Myint Myint Oo, Kyaw Zay Ya, Kyaw Zin Aung, Myint Thien, Naw May Lay Thant, Gumring Jungkum, Kyaw Thin Latt, Yinhtan Syin Bay, and Sein Day Li.

My fieldwork has been generously supported by several foundations and institutions. In particular, I would like to thank the John D. and Catherine T. MacArthur Foundation for funding research in West Kalimantan, Indonesia, Papua New Guinea, and Vietnam; the Overbrook Foundation for supporting my conservation efforts in Mexico and Brazil; the Committee for Research and Exploration of the National Geographic Society for sponsoring fieldwork in Myanmar and Vietnam; the U.S. Agency for International Development Environment Program for funding some of my work in Indonesia; the National Science Foundation for funding conservation planning in the Indo-Burma region; blue moon fund, and the Leona M. and Harry B. Helmsley Charitable Trust for supporting the community-based resource management in Myanmar and Madagascar; IKEA for funding the sustainable use of rattan in Laos and Cambodia, and the Center for Environmental Research and Conservation (CERC) at Columbia University for supporting projects in Papua New Guinea and Myanmar.

I gratefully acknowledge numerous local institutions that have hosted me and provided logistical support: Instituto Nacional de Investigación Sobre Recursos Bioticos, Grupo de Estudios Ambientales, A.C., and Instituto de Ecología in Mexico; Instituto de Investigaciones de al Amazonia Peruana in Peru; Instituto de Pesquisa Ambiental da Amâzonia in Brazil;

Minzu University in China; the World Wide Fund for Nature Greater Mekong Program in Laos and Cambodia; the Vietnamese Academy of Science and Technology; the Wildlife Conservation Society Myanmar Program; and Lembaga Ilmu Pengetahuan Indonesia.

For academic training, employment, and scientific collaboration, sincere thanks to the Forestry School at the University of Arkansas at Monticello, Hank Chamberlain, Tim Ku, and Charles Lee; the Yale School of Forestry and Environmental Studies, Tom Siccama, Mark Ashton, Rob Mendelsohn, Garth Voigt, Michael Dove, and especially my doctoral committee: F. H. Bormann, Barry Tomlinson, and Arturo Gómez-Pompa; the New York Botanical Garden, Ghillean T. Prance, Christine Padoch, Michael J. Balick, and Andrew Henderson; and the Department of Ecology, Evolution and Environmental Science at Columbia University.

A special thank you to Jion Susan Postal of the Empty Hand Zen Center for showing me a new way of looking at things. A warm nod to *Mu* and Susan Murphy Roshi at Zen Open Circle.

The majority of this book was written during a four-week writing fellowship at the Rockefeller Foundation Bellagio Center in northern Italy; bows of gratitude to Pilar Palacia and her staff. Jean E. Thomson Black at Yale University Press and Brian Boom and Michael Brown at NYBG Press provided encouragement, editorial support (and otherwise), and sage advice throughout the publication process; the careful edits of Susan Laity at Yale University Press greatly improved the manuscript. Thank you to both of these fine institutions for working together to turn my manuscript into a book—and for being an important part of my life for so many years. Elizabeth Kiernan of the New York Botanical Garden GIS Laboratory helped produce the maps.

And, finally, warm thanks and big hugs to Elysa Hammond, for her continual support and for letting me go all over the world and interact with people and plants; and to Case, Luke, and Amy for making me proud.

Index

Plants are indexed by both local name and binomial. Page numbers in *italics* refer to illustrations.

Index